D1084178

COMMUNICATION WITH EXTRATERRESTRIAL INTELLIGENCE (CETI)

THE MIT PRESS
Cambridge, Massachusetts, and London, England

COMMUNICATION WITH EXTRATERRESTRIAL INTELLIGENCE (CETI)

Edited by Carl Sagan

Copyright © 1973 by
The Massachusetts Institute of Technology

This book was set in linotype Times Roman,
printed on Nashoba Antique
and bound in Columbia Millbank Vellum MBV-4076
by The Colonial Press Inc.
in the United States of America.

Library of Congress Cataloging in Publication Data
Main entry under title:

Communication with extraterrestrial intelligence
(CETI)

English-language version of proceedings of a conference held at the
Byurakan Astrophysical Observatory, Yerevan, USSR, Sept. 5–11, 1971,
sponsored by the U. S. National Academy of Sciences and Akademiiâ
nauk SSSR.
1. Life on other planets—Congresses. 2. Interstellar communication
—Congresses. I. Sagan, Carl, ed. II. National Academy of Sciences,
Washington, D. C. III. Akademiiâ nauk SSSR.
QB54.C66   001.5'0999   73–13999
ISBN 0–262–19106–7 (hardcover)
ISBN 0–262–69037–3 (paperback)

Pub NO charge 1974/75

PREFACE

The processes of biological evolution, over a period of more than two billion years, gave rise to the creature *Homo sapiens*. As his name implies, his primary characteristic, distinguishing him from all other creatures, is the "intelligence" made possible by a brain consisting of about ten billion neurons, each of which makes dozens, even hundreds, of functional contacts with other neurons. The biological evolution of man since the appearance of *Homo sapiens* has essentially ceased, but his technological evolution has proceeded apace, extending his motor capabilities from running and climbing to submarines, jet aircraft and space capsules, from a club to giant forges and electromagnets as well as micromanipulators and the manufacture of micro-integrated circuits; extending his communicational capability from grunts to language, the printing press and the television screen; while his sensory ability was broadened to encompass the full electromagnetic spectrum, to perception of events on a scale which extends from those within the atomic nucleus to the vast reaches of the cosmos. By these means, man became the most adaptable of all earthly species, successfully dwelling in an econiche which is the entire planet. His brain, the physical structure of which was completed with Cromagnon man, has not evolved further, but it too has acquired powerful external accessories, most notably memory in the form of the printed record and both the computational and data processing assistance of the "computers" it invented. It is, perhaps, rueful to consider that these very accomplishments of his brain, of themselves, have eliminated those pressures which might otherwise have fostered further his biological evolution!

Man has now come far along the trail to understanding himself and where he is in the physical universe. And, for the first time, he recognizes that he has, or may be able to devise, the tools to begin to examine whether there are other equally or more intelligent beings somewhere in that universe. The conference here reported is significant in part for what it accomplished in that regard but primarily for the historical fact of its occurrence, a punctuation mark in the history of mankind.

Philip Handler, President
National Academy of Sciences
Washington, D.C.

INTRODUCTION

This is the proceedings of a remarkable conference on a subject which, only a few years ago, would have been considered an unthinkable topic for serious scientific discussion: Communication with Extraterrestrial Intelligence (CETI). But our current knowledge of the physical and biological sciences makes it appear, at least to many of us, that the chances of there being extraterrestrial intelligence is much greater than scientists thought possible a few decades ago. And what is beyond doubt is that the level of technological advance on the planet Earth in the last few decades has been so spectacular that the means for contact with extraterrestrial intelligence are potentially within our hands. Unless or until contact is actually established, the subject is destined to be riddled with uncertainty and honest differences of opinion. But I think no serious student of the subject questions its significance—both for science and for the deepest philosophical and human questions. In a very real sense this search for extraterrestrial intelligence is a search for a cosmic context for mankind, a search for who we are, where we have come from and what possibilities there are for our future—in a universe vaster both in extent and duration than our forefathers ever dreamed of.

The concept of this conference originated in discussions held from about 1967 onward in a variety of cities throughout the world, including Moscow, Prague, New York and Washington. The initial discussants were N. S. Kardashev and I. S. Shklovsky in the Soviet Union; Philip Morrison, Frank Drake, and myself in the United States; and Rudolf Pešek, in Czechoslovakia. Professor Pešek, the International Astronautical Federation, and the Czech Academy of Sciences made an early offer to host this meeting. While events led in another direction, we are grateful to them for their important early efforts to help concretize this meeting. The conference as it actually materialized was sponsored jointly by the National Academy of Sciences of the United States of America and the Academy of Sciences of the Union of Soviet Socialist Republics. Support for U.S. participants was provided by the National Science Foundation. In the United States Harrison Brown, the foreign secretary of the National Academy, Robert Forcey and Jesse Mitchell played important roles in the meeting arrangements. The support and encouragement of Philip Handler, President of the National Academy of Sciences, is gratefully acknowledged. The hospitality of our Soviet hosts was extremely gracious from the arrival of most of us in Moscow to the departure of most of us from Moscow; and in the Armenian Soviet Socialist Republic the hospitality of Academician V. A. Ambartsumian and his staff was instrumental for the success of the meeting. G. S. Khromov, Vice President of the Astronomical Council of the USSR Academy of Sciences, helped in expediting the return of corrected manuscripts from Soviet participants.

The stenotypic transcript of the conference—the fundamental base material on which the published proceedings are based—

was made by Mrs. Floy Swanson, whose care and scrupulous
attention to detail cannot be praised enough. Mrs. Swanson's
participation was made possible by a special fund provided by
Doctor Hans Mark, Director of the Ames Research Center, National Aeronautics and Space Administration. Without Doctor
Mark's timely assistance, these Proceedings would not have been
possible. Mrs. Phyllis Morrison controlled an elaborate multi-
input magnetic tape recording system and is also responsible for
most of the candid photographs displayed in this volume. We
are very grateful to Mrs. Morrison for major contributions to
the success of the conference.

I think it does no other participant an injustice to say that the
keystone of the success of the meeting was Boris Belitsky, science editor for English language programming on Radio Moscow. In the experience of all conference participants I have
talked to, Mr. Belitsky is, by a large factor, the most articulate,
adroit and competent *viva voce* scientific translator in both Eng-
lish/Russian and Russian/English any of us have experienced.
Mr. Belitsky's real time translations in both directions were flaw-
less and scientifically literate. Indeed there were several instances where he corrected stylistic or scientific infelicities in
both languages. In order to undertake meaningful communication with extraterrestrial intelligence it would seem that meaningful communication among terrestrial intelligence is a prerequisite. The goodwill of the participants and Mr. Belitsky's
efforts made possible such terrestrial exchanges among individuals of different backgrounds, nationalities, languages, and scientific disciplines.

The meetings were held in the main conference room of the
Byurakan Astrophysical Observatory of the Armenian Academy
of Sciences. The Soviet and American Organizing Committees
and a few other participants lodged at the Observatory; the other
participants, in nearby Yerevan, the capital of the Armenian

SSR. We were all in daily sight of Mount Ararat, on which Noah's Ark is said to be beached. Engaging personal accounts of the nonscientific (and a little of the scientific) aspects of this meeting have been published by Freeman Dyson in the *New Yorker* November 6, 1971, p. 126; and by William H. McNeill in the *Chicago Magazine* May/June, 1972. Numerous articles about the meeting have also been published in the Soviet press.

The form of the discussions was that of an initial presentation of a subtopic by a discussion leader and then an often lively and vigorous range of comments, criticisms, and free association. Chairmen of the various sessions were Ambartsumian, Crick, Gold, Kaplan, McNeill, Orgel, Pešek, Sagan, Shklovsky, and Troitsky. The sequence of topics, ranging from the very speculative to the very concrete, are given in the table of contents and follow closely the actual order of discussion at the meeting. In the interest of clarity some discussions have been relocated to make their antecedents clearer. With minor exceptions all the content of the meeting is included here. Because of the large spectrum of human knowledge relevant to the topic of CETI and the enthusiasm of the participants, some sessions ran from early morning to quite late at night, causing more than usual strain on the participants, stenotypist, and translators.

A set of recommendations and conclusions from the conference was issued as a communiqué, and was worked on over many hours by a small group which included Doctors Kardashev, Shklovsky, Gindilis, Troitsky, and Kaplan from the Soviet Union and Drake, Morrison, and myself from the United States. Mr. Belitsky played an active role in these discussions. The draft resolution of the organizing committees was subjected to a sentence-by-sentence discussion in a plenary meeting at the last day of the conference, in which all participants present had veto power. The communiqué reproduced here therefore represents a view which all participants could agree to—although there was a large majority willing to vote for a stronger statement on the likelihood of extraterrestrial intelligence. The communiqué, in a way a summary of the conference proceedings and in a way an aperture to future studies on the problem of CETI, is at the end of this volume.

The authors of major discussions have generally had an opportunity to check the transcript for factual errors or errors of transcription, but every effort has been made to reproduce closely the character of the actual discussions. The flavor of this scientific debate is, I think, almost as interesting as the subject matter itself. And having read the text many times I still find places where the flow and brilliance of new ideas among the participants is electrifying.

Because this is a verbatim record of an interdisciplinary scientific discussion embracing many diverse subjects, it may not be entirely easy reading for the layman. Nevertheless—perhaps in part because many scientists are moderately illiterate in areas not their speciality—I believe the major part of the flow and exchange of ideas can be understood readily by laymen interested in science but with no particular technical background. Some discussions, for example the account by Richard Lee of the evolution of man and that by Kent Flannery on the origins of civilizations, stand alone as extraordinarily able short summaries, understandable to everyone, of important and difficult subjects. To aid the layman there is a glossary of technical terms appended at the end of the book and an extensive index to aid in the cross-connection of ideas. Some concepts, such as Kardashev's classification of extraterrestrial civilizations as the successively more advanced Types I, II, and III, or Hans Freudenthal's language *Lincos* for interstellar discourse, are introduced by the participants without any explanatory material. Likewise some of the discussions of the details of interstellar radio communication assume prior knowledge on the part of the reader. Perhaps I may be excused if I say that some introduction to all of these ideas can be garnered from the book *Intelligent Life in the Universe* by I. S. Shklovskii and Carl Sagan (San Francisco: Holden-Day, 1966; New York: Dell Publishing Company, 1967 and many later printings). Other relevant books on the subject are *Interstellar Communication,* edited by A. G. W. Cameron (New York: Benjamin, 1965); *We Are Not Alone,* by Walter Sullivan (New York: McGraw-Hill, 1964); *Extraterrestrial Civilizations,* edited by G. M. Tovmasyan (Erevan: Armenian Academy of Sciences; English translation by Z. Lerman, Israel Program for Scientific Translations, IPST 1823, 1967); *Extraterrestrial Civilizations: Problems of Interstellar Communication,* edited by S. A. Kaplan (in Russian, Moscow, 1969; English translation by Israel Program for Scientific Translations, IPST 5780, NASA Technical Translation TT F-631, 1971); and *The Cosmic Connection,* by Carl Sagan (New York: Doubleday, 1973). The Tovmasyan book is the proceedings of a prior conference held on this subject in the Soviet Union. A summary of the first conference on extraterrestrial intelligence, held in the United States in 1961, can be found in the first three references above.

While many of the participants are very well-known in the scientific community, the layman may not be familiar with most of them. Therefore a list of participants follows this Introduction. I here append a few words about a perhaps representative sampling of the participants. Philip Morrison along with G. Cocconi made the first proposal that radio waves of wavelength 21 cm, at which interstellar hydrogen likes to emit and absorb, are being used for interstellar communication beamed at the earth. Frank Drake, in Project Ozma, performed the first (unsuccessful) such search. V. S. Troitsky reports in this volume a more elaborate set of successor experiments to Ozma in searching for signals from extraterrestrial civilizations. I. S. Shklovsky and N. S. Kardashev have played leading roles in setting the theoretical and intellectual climate for such searches in the Soviet Union. Charles Townes made the first proposal for use of lasers in interstellar communications; Townes received the Nobel Prize in physics in 1964 for inventing the maser and laser. Francis Crick, another skeptical Nobel laureate, is best known for his discovery, along with J. D. Watson, of the double helix structure of DNA and for his specification of the nature of the genetic code. Shklovsky, V. L. Ginzburg, and Thomas Gold are among the world's leading theoretical astrophysicists. Freeman Dyson, a mathematician and theoretical physicist, was one of the first to call attention to the possible astroengineering activities of very advanced civilizations—an expected large-scale reworking of their local cosmic environments. M. Ya. Marov, Y. K. Khodarev, and V. I. Moroz play leading roles in Soviet planetary studies. Richard Lee has studied both nonhuman primates and the !Kung bushmen, with whom he lived for several years, in the Kalahari Desert. Kent Flannery is one of the few archeologists in the world to have made important studies both in Meso-America and in the Near East. Y. N. Pariisky is the director of a new Soviet Radio telescope being built in the Caucasus Mountains, which will be devoted in part to CETI. W. H. McNeill is a leading American historian, the author of the best selling *The Rise of the West*. L. Mukhin is responsible for Soviet exobiological instrument development. Marvin Minsky is a leading designer of machines which have attributes we call intelligent. V. A. Ambartsumian is President of the Armenian Academy of Sciences, Director of the Byurakan Observatory, and a member of the Supreme Soviet. David Heeschen is the Director of the National Radio Astronomy Observatory, while Drake is the Director of the National Astronomy and Ionosphere Center, the two facilities in the United States with the largest steerable radio antennas. Bernard Burke is co-discoverer of the radio burst emission from the planet Jupiter. B. V. Sukhotin is a leading expert on the decripting of messages. Leslie Orgel has performed important experiments in prebiological organic chemistry and also originated one of the principal theories on the nature of aging in living systems. S. A. Kaplan is an expert on the interstellar medium. B. M. Oliver is an electrical engineer who directed a large-scale study of a proposed major program to search for extraterrestrial intelligence which he describes in this volume. John Platt is a physicist primarily concerned with the human condition and how to improve it. G. M. Tovmasyan is a radio astronomer who edited the proceedings of the first Soviet conference on CETI. This short list is intended only to convey something of the breadth and depth of the conference participants. The American representation was proportionally somewhat stronger in the biological and social sciences than the Soviet delegation. The Soviet delegation was correspondingly stronger on the practical astronomical side. But the distribution of disciplines appears to have been mutually complementary.

I am indebted to Doctors Terrence Fine, G. G. Simpson, James Elliot, Martin Harwit, and Joshua Lederberg, who, while unable to attend the conference, nevertheless sent along short contributions which are printed in the appendices; and to the Master of Birkbeck College for permission to print here Freeman Dyson's remarkable essay, "The World, the Flesh, and the Devil," which touches on so many of the topics of the present volume. I am very grateful to Mary Szymanski, Carol Smith, Marye Wanlass, and JoAnn Cowan for retyping the manuscript of these proceedings. Their time was supported in part by the National Aeronautics and Space Administration. These Proceedings will also be published in a Russian language edition by the Publishing House "Mir" in Moscow under the editorship of S. A. Kaplan.

Carl Sagan
Ithaca, New York   9 November 1972

PARTICIPANTS

Soviet Organizing Committee

V. A. AMBARTSUMIAN, Chairman
Byurakan Astrophysical Observatory, Armenian Academy of
Sciences.

N. S. KARDASHEV
Institute for Cosmic Research, Soviet Academy of Sciences,
Moscow.

I. S. SHKLOVSKY
Institute for Cosmic Research, Soviet Academy of Sciences,
Moscow; and Shternberg Astronomical Institute, Moscow State
University.

V. S. TROITSKY
Institute for Radiophysics, Gorky State University, Gorky.

U.S. Organizing Committee

C. SAGAN, Chairman
Center for Radiophysics and Space Research, Cornell University,
Ithaca, New York.

F. D. DRAKE
Center for Radiophysics and Space Research, Cornell University, Ithaca, New York.

P. MORRISON
Department of Physics, Massachusetts Institute of Technology, Cambridge, Massachusetts.

Soviet Participants

S. Y. BRAUDE
Institute of Radiophysics and Electronics, Ukrainian Academy of Sciences, Kharkov.

E. M. DEBAI
Shternberg Astronomical Institute, Moscow State University.

L. M. GINDILIS
Shternberg Astronomical Institute, Moscow State University.

V. L. GINZBURG
Lebedev Physical Institute, Soviet Academy of Sciences, Moscow.

G. M. IDLIS
Astrophysical Institute, Kazakh Academy of Sciences, Alma-Ata.

S. A. KAPLAN
Institute for Radiophysics, Gorky State University, Gorky.

V. V. KAZUTINSKY
Institute of Philosophy, Soviet Academy of Sciences, Moscow.

E. E. KHACHIKYAN
Byurakan Astrophysical Observatory, Armenian Academy of Sciences, Erevan.

Y. K. KHODAREV
Institute for Cosmic Research, Soviet Academy of Sciences, Moscow.

Y. I. KUZNETZOV
Institute of Energetics, Moscow.

B. E. MARKARIAN
Byurakan Astrophysical Observatory, Armenian Academy of
Sciences, Erevan.

E. S. MARKARIAN
Institute of Philosophy and Law, Armenian Academy of
Sciences, Erevan.

M. Y. MAROV
Institute of Applied Mathematics, Soviet Academy of Sciences,
Moscow.

E. MIRZABEKIAN
Institute of Radiophysics and Electronics, Armenian Academy
of Sciences, Erevan.

L. V. MIRZOYAN
Byurakan Astrophysical Observatory, Armenian Academy of
Sciences, Erevan.

V. I. MOROZ
Institute for Cosmic Research, Soviet Academy of Sciences,
Moscow, and Moscow State University.

L. M. MUKHIN
Institute for Cosmic Research, Soviet Academy of Sciences,
Moscow.

L. M. OZERNOY
Lebedev Physical Institute, Soviet Academy of Sciences,
Moscow.

B. I. PANOVKIN
Radioastronomical Council, Soviet Academy of Sciences,
Moscow.

Y. N. PARIISKY
Special Astrophysical Observatory, Soviet Academy of Sciences,
Leningrad.

N. T. PETROVICH
Electrotechnical and Communications Institute, Moscow.

R. G. PODOLNY
"Knowledge is Power," Moscow.

V. A. SANAMYAN
Byurakan Astrophysical Observatory, Armenian Academy of Sciences, Erevan.

V. I. SIFOROV
Institute for Information Transmission, Soviet Academy of Sciences, Moscow.

V. I. SLYSH
Institute for Cosmic Research, Soviet Academy of Sciences, Moscow.

B. V. SUKHOTIN
Institute of the Russian Language, Soviet Academy of Sciences, Moscow.

M. L. TER-MIKAELIAN
Byurakan Astrophysical Observatory, Armenian Academy of Sciences, Erevan.

G. M. TOVMASYAN
Byurakan Astrophysical Observatory, Armenian Academy of Sciences, Erevan.

U.S. Participants

B. BURKE
Department of Physics, Massachusetts Institute of Technology, Cambridge, Massachusetts.

F. DYSON
Institute for Advanced Study, Princeton, New Jersey.

K. FLANNERY
Department of Anthropology, University of Michigan, Ann Arbor.

T. GOLD
Center for Radiophysics and Space Research, Cornell University, Ithaca, New York.

D. HEESCHEN
National Radio Astronomy Observatory, Charlottesville,
Virginia.

S. VON HOERNER
National Radio Astronomy Observatory, Charlottesville,
Virginia.

D. HUBEL
Department of Physiology, Harvard University Medical School,
Cambridge, Massachusetts.

K. KELLERMANN
National Radio Astronomy Observatory, Charlottesville,
Virginia.

W. H. MCNEILL
Department of History, University of Chicago, Chicago, Illinois.

M. MINSKY
Department of Electrical Engineering, Massachusetts Institute
of Technology, Cambridge, Massachusetts.

B. M. OLIVER
Hewlett-Packard Corporation and Stanford University, Palo
Alto, California.

L. ORGEL
The Salk Institute, La Jolla, California, and London University,
London, U.K.

J. R. PLATT
Michigan Mental Health Center, University of Michigan, Ann
Arbor.

G. STENT
Department of Virology, University of California, Berkeley.

C. TOWNES
Department of Physics, University of California, Berkeley.

Participants from Other Nations

F. H. C. CRICK
Medical Research Council Laboratory of Molecular Biology,
Cambridge University, U.K.

R. B. LEE
Department of Anthropology, University of Toronto, Toronto,
Ontario, Canada.

G. MARX
Department of Physics, Budapest University, Budapest,
Hungary.

R. PEŠEK
Czech Academy of Sciences, Prague, Czechoslovakia.

V. A. Ambartsumian

Carl Sagan

All the photographs
except that of Frank
Drake were taken by
Phyllis Morrison at the
CETI conference.

Boris Belitsky doing
simultaneous transla-
tion

Thomas Gold

Thomas Gold discuss-
ing the angular
momentum of stars

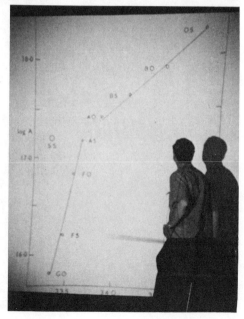

S. A. Kaplan, left, translates for Kent Flannery

David Hubel describes the neurology of the hand

Richard Lee

W. H. McNeill in a
pensive moment

Leslie Orgel

Francis Crick makes a
point to Marvin
Minsky

L. M. Gindilis illus-
trates a point

E. S. Markarian reading the text of a presentation

Comments by R. G. Podolny

Carl Sagan at the left
and I. S. Shklovsky,
right, unimpressed by
an argument

Remarks by G. M.
Idlis

Marvin Minsky demonstrates a two-stage water-propelled toy rocket at the corner of a pool at the Byurakan Observatory

Y. K. Khodarev (dark shirt, bottom) Assistant Director of the Institute for Cosmic Research, Soviet Academy of Sciences, Moscow, and R. Pešek (jacket) of the International Academy of Astronautics and Sebastian von Hoerner (in front of Pešek) who has written on interstellar spaceflight, watch this experimental effort

There was a second stage failure and the mission aborted

Freeman Dyson

N. S. Kardashev

V. L. Ginzburg

Frank Drake

Y. N. Pariisky

V. S. Troitsky tells about Soviet efforts to listen in on interstellar radio signals

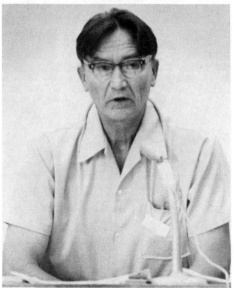

V. I. Moroz presents
some calculations at
the blackboard

B. M. Oliver discusses
Project Cyclops

Charles Townes at left
and V. L. Ginzburg at
right

N. T. Petrovich

B. I. Panovkin

I. S. Shklovsky, right, making a point with some vigor to Philip Morrison

PROSPECT

$$N = R_* f_p n_e f_l f_i f_c L$$

AMBARTSUMIAN: The Organizing Committee has asked me to make a few introductory remarks at the opening of this Conference on Communication with Extraterrestrial Intelligence (CETI). Our conference has been convened by the Academy of Sciences of the USSR and the National Academy of Sciences of the United States. Accordingly, it is a Soviet-American Conference. The Armenian Academy of Sciences offered to hold the Conference at Byurakan and this offer was accepted. Accordingly, the Armenian Academy has played a considerable part in the organizing work.

  The Organizing Committee decided that while keeping the conference bilingual, it would be expedient to send personal invitations to a small number of selected scientists from other countries at the same time and it is thanks to this that we have the privilege of seeing in our midst Doctor Crick from the United Kingdom, Doctor Marx from Hungary, and Doctor Pešek from Czechoslovakia. Their participation will undoubtedly be very useful and valuable to us.

Both the initiators of the Conference and the organizers are fully aware of the fact that the subject matter chosen contains much that is nebulous. One thing, however, is clear. It is that the conference is broadly interdisciplinary by nature. For this reason, it was clear that the discussion of the subject at this naturally preliminary stage required that many different sciences be represented. Though we could not assure a sufficiently uniform representation of all the sciences relevant to CETI, we are happy that scientists representing a fairly wide range of problems are present.

Of course, there are some who may feel that a discussion of extraterrestrial intelligence and communication with such intelligence is somewhat premature, since there is as yet no direct evidence of the existence of such intelligence. But the sponsors of the conference are of the opinion that this evidence must be sought actively and that the problem must be studied fundamentally, drawing upon all the data available to contemporary astronomy, planetary sciences, biology, and social sciences.

How close have we come to a proper scientific presentation of the problem to justify the holding of so representative a conference? It is my conviction that there are weighty grounds for undertaking a profound study of the problem before us. In this context, one may point to the progress made in radio astronomy, and especially in the techniques of receiving faint signals from distant cosmic sources. There have been tremendous advances in these fields and I believe that the outlook for further strides is improving all the time.

On the other hand, contemporary astronomy has come close to becoming all-wave astronomy. This increases our capacity for receiving possible signals from extraterrestrial intelligence (ETI). Advances in various areas of cybernetics and information theory should make it possible to tackle the problem of the possible CETI communications channels and the modes of instruction that could be employed by superior civilizations in addressing other civilizations in the absence of any prior and explicit understanding.

Contemporary biology has revealed some of the mysteries of the life processes. We are on the way to being able to analyze the various forms of life and even the various mechanisms whereby intelligence asserts itself. Space exploration has added to our understanding of the moon and the planets. We are on the way to gaining an understanding of their origins. All this has produced a situation in which a discussion on CETI may prove sufficiently fruitful. Therefore, this conference seems quite justified. Even if our discussion merely gives us a better understanding of which approaches to adopt, the time we use will not have been lost.

Prior to this conference rather successful similar conferences have been held. Our American colleagues of the National Academy of Sciences held a conference on CETI in 1961. Also, seven years ago Soviet scientists met here at Byurakan for their first conference on the subject. The proceedings of the Byurakan Conference were published in Russian and this book was afterwards translated into English.

This first conference made it evident that discussion of this subject among scientists in different fields was useful, and we know that in other countries, too, there is a keen interest in this subject. I think, therefore, that Professor Shklovsky was right when he said to me that before we are able to solve the problem of communicating with extraterrestrial civilizations it might be a good thing for there to be communication on the subject among nations, and that is precisely the purpose of our conference.

We all realize that the discovery of the first ETI could have colossal implications for man's advancing knowledge, of the same order of magnitude as the launching of the first Sputnik or the harnessing of atomic energy, even if not so instantaneous an

event and, therefore, not so spectacular. Nevertheless, since such
a discovery may prove so important, so colossally important,
most of the participants in the conference are very enthusiastic
about the problem; our guests from across the ocean did not
shrink from their lengthy and difficult journey.

Permit me to welcome all of you who have come here for this
conference, all the participants. It will probably be expedient at
the end of our work to draw up and approve brief conclusions
as to the present state of the problem. It would also seem desirable that at the end of the conference we release an agreed-
upon communiqué, so let me add one technical remark. We have
facilities for translation here. We have the capacity for simultaneous translation but it was decided at the Organizing Committee's meeting that the discussion initiators, at any rate, would at
first be translated consecutively so as to make for a more accurate translation and to assist the subsequent discussion, and
that will make it easier for all of us. It will also be easier for
those of us who are able to follow both languages.

Permit me, then, to conclude by wishing all of you success in
our work.

I would now like to call upon Doctor Sagan of the United
States for his introductory remarks.

SAGAN: Thank you, Academician Ambartsumian. We very much
appreciate your kind hospitality and this elegant meeting room.
The participation of United States delegates has been made
possible by support from the National Academy of Sciences and
the National Science Foundation of the United States, and the
support for Mrs. Swanson, who is making a stenotypic transcript of all the remarks of this meeting for possible publication,
has been paid for by the National Aeronautics and Space Administration.

The word CETI which has been devised for this meeting is
I think appropriate in three different respects. First, it is an

acronym for Communication with Extraterrestrial Intelligence. Second, it is the Latin genitive for whale, which is of some interest to this discussion; the cetaceans are undoubtedly another intelligent species inhabiting our planet, and it has been argued that if we cannot communicate with them we should not be able to communicate with extraterrestrial civilizations. And finally, one of the two stars which was first examined by Frank Drake in Project Ozma, the first experimental undertaking along those lines, was Tau Ceti.

It has been suggested that, in order to set a coherent framework for some of the following discussion, we should express one formulation for the number of technical civilizations in the Milky Way Galaxy at or beyond our level of technological advance. As you will see in the program, the first two and a half days of this conference are devoted to a discussion of these constituent factors. There are many formulations of this problem. In the simplest formulation due in its original form to Drake, N, the number of such extant civilizations in the Galaxy, can be written as the product of seven factors:

$$N = R_* f_p n_e f_l f_i f_c L. \tag{1}$$

Here R_* is the rate of star formation averaged over the lifetime of the Galaxy, in units of number of stars per year; f_p is the fraction of stars which have planetary systems; n_e is the mean number of planets within such planetary systems which are ecologically suitable for life; f_l is the fraction of such planets on which the origin of life actually occurs; f_i is the fraction of such planets on which, after the origin of life, intelligence in some form arises; f_c is the fraction of such planets in which the intelligent beings develop to a communicative phase; and L is the mean lifetime of such technical civilizations.

The factor R_* is the province of astrophysics, as is f_p; n_e is determined at the boundary between astronomy and biology; f_l

is largely a topic of organic chemistry and biochemistry; f_i is a topic in neurophysiology and the evolution of advanced organisms; f_c is a topic in anthropology, archaeology, and history. And the last term is in the very nebulous area of predicting the future of societies; it involves psychology and psychopathology, history, politics, sociology, and many other fields. The reliability of estimates declines markedly from R_* to L in equation (1). But, quite apart from how well we can estimate these factors, it is remarkable that there exists a single problem that involves so intimately subjects ranging from astrophysics and molecular biology to archaeology and politics.

The final remark I wish to make concerns the nature of these constituent factors. The question of what is meant by a probability in this context has been raised several times. I would like to stress that the concept of probability in this equation changes as one goes from R_* to L in equation (1). The question of the number of stars formed in unit time in the Galaxy is a statistical quantity. One determines it by counting and by some theoretical considerations. There are $\sim 10^{11}$ stars in the Galaxy, which is $\sim 10^{10}$ years old; $R_* \sim 10$ stars/year is the only factor even approximately well determined. But as we go on, we encounter the problem of extrapolating from a few or from only one example, plus other relevant information. Finally, as we get to L, we have reached a point where, fortunately for us but unfortunately for the discussion, there is not even a single example of the lifetime of a technical civilization. We have limits but no absolute value for L even for Earth.

So, what is meant by assessing a probability for the likelihood of development of intelligence or the likelihood of development of a communicative phase? I would like to refer to a paper prepared for this meeting (see Appendix A) by Professor Terrence Fine of Cornell University, who has just written a book called, *Theories of Probability: An Examination of Foundations.* He points out that in addition to enumerative probability estimates and probability estimates made because one thinks he has a firm understanding of the underlying physics, there is a third kind of probability estimate, subjective probability. And that, I am afraid, is where we are at. Fine writes:

The subjective or personalistic interpretation of probability maintains that probability estimates are derived through a largely unassisted process of introspection and are then applied to the selection of optimal decisions or acts, such as the allocation of research resources. The subjective view boldly admits of the subjective elements in most other concepts of probability and even encourages the holder to fully use his informal judgment, beliefs, experience in arriving at probability estimates whose objective is decision making and the interpersonal communication of individual judgment, and not, say, assessing the "truth" of propositions.

While subjective probability statements are indeed personal, they are not arbitrary. There are reasonable axioms of internal consistency between assessments and constraints that force the user to learn from experience in a reasonably explicit way. While it is not possible to criticize any single subjective probability assessment, it is possible to criticize a collection of such assessments.

And then Fine concludes with the following words:

The subjective approach has been widely discussed and applied in management decision making and in reliability analysis. While it has evident limitations, so do all of the other approaches to probability. We would judge that the concept of subjective probability is at present the only basis upon which probability estimates can be made about extraterrestrial intelligent life.

EXTRASOLAR PLANETARY SYSTEMS

f_p, n_e

GOLD: It is my task to discuss the solar system and the likelihood that other such systems occur elsewhere in the Galaxy. We are interested, of course, because this solar system is the one place where we know that life with some degree of intelligence has occurred. We may, of course, be quite mistaken in thinking that solar systems are particularly favorable sites for life, and it may be that we should investigate other nonplanetary locales for finding intelligent life. But we believe we have reasons not only of an egocentric nature for concentration on planets. We think we understand that the circumstances of planets are particularly favorable for the evolution of some necessary adjuncts to life.

We think we need complicated chemistry, and that fixes a certain temperature range which probably must not exceed by very much the temperatures we have on the earth. We also think that the temperatures must not be extremely low or else chemical reactions would take place so slowly that the time scale available would be insufficient for the development of any complex systems.

In addition to the temperature requirement, we have the requirement that there be a nonthermal energy flow through the system. We believe that this nonthermal energy flow had best be in the form of photons, but perhaps that is not an absolute requirement. If it is in the form of photons, they must not have too high an energy or else they will destroy the complex chemistry which we seek to create.

We see that on the earth we have these circumstances; we have the temperature range that we discussed and we have, in addition, solar photons which allow energy to be abstracted and used in biological systems.

Now we would like to understand our solar system in order to answer the following questions: How common is it in the universe? Around what types of stars are there planetary systems? What is the distribution of sizes of planets and their distances from their stars? What temperatures occur on those planets? What rotations do these planets have? What time scale is available to them for evolution?

Earlier discussions of the solar system held that it was a rather rare event that gave rise to ours. In recent times we have come upon a set of considerations, which I will present, which make it appear much more likely that the origin of the solar system was a common event. We believe now that there must have existed at one time a solar nebula, surrounding the sun; we believe it was in the neighborhood of 5 billion years ago, and we can discuss in some detail the dynamical behavior that this nebula must have possessed.

The fact that the planets are distributed in such a narrow plane implies that what started the system was a gas, because only in that way can we understand the attenuation of a sheet of matter into a very narrow plane. We then believe that it was necessary for the angular momentum of the system to be redistributed in such a manner that only approximately 0.2 percent of the mass of the system came to possess 98 percent of the angular momentum. That cannot have occurred readily in a single event and we believe that a period must have existed during which the angular momentum was gradually so redistributed. The second phase that must have taken place is a process of condensation (and I will say more about that), and the third, a loss of hydrogen.

Of the condensation we understand some features, but not all. We understand that the more refractory materials condensed nearer the sun and then progressively less refractory materials condensed at greater distances; that the terrestrial planets were formed out of materials such as silicates and iron, and then the giant planets—Jupiter and so on, lying beyond the terrestrial planets—were formed initially, apparently, out of materials such as carbon, nitrogen, and oxygen, chiefly; they later acquired a large quantity of hydrogen and helium.

As you know, Jupiter and Saturn have this very large amount of hydrogen and helium but Uranus and Neptune do not. They are probably made mostly out of carbon, nitrogen, and oxygen, and that brings me to this point: it is necessary to suppose that in a late phase of the formation of the solar system a large quantity of hydrogen was lost.

Let me say a little bit more about the condensation. Particles must have condensed out of the gas according to their vapor pressures and perhaps according to other considerations as well, such as nucleation processes and processes of crystal growth. Some would have us believe that ionization potentials played a part in the selection of the condensation process as well. At any rate, small particles were formed first, and their orbits caused them eventually to coalesce by processes that, once they started, apparently occurred quite rapidly.

It is surprising that the planets are so well spaced out, so that there is no real risk of a collision. If you started in a random fashion, accumulating particles in space, you would end up with systems in which collisions were quite common. It is therefore likely that there was a period of growth of bodies and of collisions, out of which a set of bodies which no longer were on collision orbits with each other gradually selected itself. Perhaps we see the last remains of this collision phase in the very many craters that we observe both on Mars and on the moon.

For the outer planets, the condensation process of the ice particles of carbon, nitrogen, and oxygen must also have gone through such collision events, and there the clue that this occurred may be found in the comets. Then, once enough carbon, nitrogen, and oxygen had been put together to attract hydrogen gas, the condensations grew by the gravitational accumulation of hydrogen and helium gas.

These points are now largely accepted by people who discuss the origin of the solar system; but that, of course, still leaves many debatable points on which I have not touched. The mechanism by which angular momentum was transferred is by no means clear. We think that magnetic fields are involved. The details of the accumulation processes and the condensation mechanisms, the process of hydrogen loss from the outer solar system where it must once have been present but is now clearly absent—these mechanisms are quite poorly understood.

Still, I explain all these points because they may help you, and especially the nonastronomers among you, to consider the likelihood of such systems elsewhere. We are interested in solar systems elsewhere that have a time scale as long as ours, or of that order, in order that there may be enough time to evolve intelligent life. That immediately means that we are not interested in systems that might form around the very massive, very bright stars, since those stars have very short lifetimes; their later evolution would make any further life on their planets impossible. This consideration would restrict us to stars from those a little bit brighter, a little bit more massive than the sun, down to stars very much cooler and very much less massive than the sun.

Now let me present some information from other stars relevant to extrasolar planetary systems. The most important point about these stars concerns their angular momentum. It is of great interest that stars like the sun have peculiarly small amounts of angular momentum. It is quite clear that they must have lost angular momentum since the time of their formation. If one were to take an arbitrary mass equal to the solar mass, anywhere in the Galaxy, and gather it together, it would spin very much faster than the sun; it would spin very near its bursting speed. The fact that all the very many stars that are in the mass range from slightly less massive to slightly more massive than the sun—the fact that all those stars have been observed to have small amounts of angular momentum surely must mean that there has been a process of loss. Figure 1 shows one such demonstration of the statistical situation. The massive stars are on the right, the less massive stars are on the left. The sun is a G0 star. Most stars are less massive than the sun. You see that for the massive stars there is a tendency for the angular momentum to be quite high, while for the lower-mass stars it suddenly decreases immensely. If we were to augment the angular momentum per unit mass of the sun by the addition of the planetary system, then the point for the sun in Figure 1 would fall almost on the same curve as the more massive stars do.

This is the strongest point that we have for supposing that the formation of the solar system was the means by which this special loss of angular momentum occurred. I am saying it is the strongest argument we have. I am not asking you to accept it, but it is worth discussing. It may, of course, be that the angular

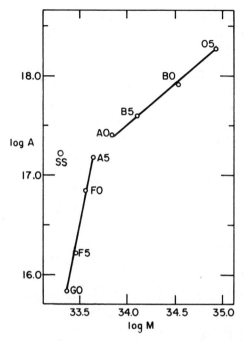

Figure 1. The logarithm of the angular momentum A of stars versus the logarithm of the masses M of stars. The curve shows a striking break in early A spectral types. It is possible that the stars at the left end portion of the diagram have transferred their initial angular momentum to planetary systems about them. The point marked SS shows where the sun would lie if the angular momentum of the solar system were incorporated into the rotational angular momentum of the sun.

momentum is lost in a process that does not always lead to the formation of a solar system, and indeed this is probable: the amount of hydrogen that appears to have been lost from the outer part of the solar system would have contained much more angular momentum than all the rest of the solar system together. If that could be lost in our case, you might well worry whether in other cases it was not the entire amount of angular momentum that was so lost. Since we don't know the process involved here, we cannot readily judge whether it was a likely one or an unlikely one.

Now I want to discuss briefly the probability of the formation of planets approximately like our planets, assuming that we start with a solar nebula. Dole* has calculated with a computer the kind of planetary systems that would be formed from random accumulation within the nebula. I will present some examples from the many such calculations that he has done, just to show that it is not exceedingly unlikely that such processes can lead to systems like ours.

Figure 2 is our actual solar system. The size of the circle indicates the mass of the planet. The trident underneath indicates the eccentricity of the planetary orbit. The distance from the left-hand side of the figure is the distance from the sun. This is our actual solar system; I will now present some examples of random nebular accumulations, and you can judge whether you think that ours might be one such random example.

In Figure 3, the eccentricities are on the whole a little bit larger than in our solar system. The planets would be a little more in danger of collision. I imagine that if these calculations were carried out for a lot longer so that collisions were allowed to occur followed by a period of scattering and reaccumulation, the computer simulation would probably resemble even more closely our actual solar system.

In Figure 4 is another set of such calculations. As you see, by and large, you obtain an approximation to Bode's Law, as it is called, the approximate distribution of planets from the sun. You tend to make the large planets in the right place. So I don't think that the condensation process is really a great mystery anymore.

* S. Dole, Computer simulation of the formation of planetary systems. *Icarus* 13 (1970): 494–508.

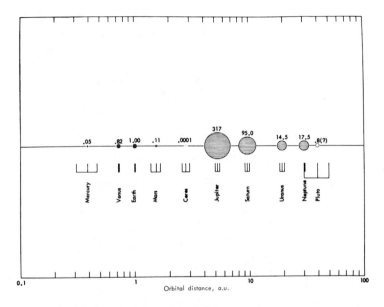

Figure 2. A representation of the solar system. The distances of the planets from the sun are shown in astronomical units (a.u.). The earth is 1 astronomical unit from the sun. The masses of the planets are shown compared to the mass of Earth, taken here as one unit. [From S. Dole, *Icarus* 13(1970): 504.]

But let me emphasize that there is one thing that still is quite unexplained by any of this discussion, and that is the occurrence of the thirty-one or more natural satellites that we have in our solar system. The satellites require some special explanation. The theories of origin of our moon have been specific to our moon for the most part, and that is not satisfactory when you realize that there are another thirty moons to be explained. But I have no doubt that this would not change the basic discussion that I have given you, and that we will merely find in the course of time, I trust, what the detailed physical process has been that has given rise to the high probability of forming satellite systems.

I now come to what we have seen in the vicinity of other stars

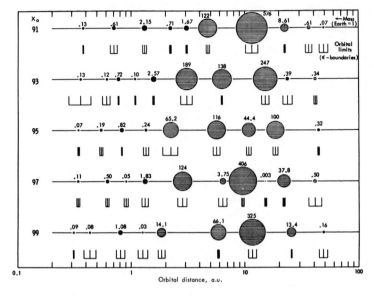

Figure 3. An end product of a computer simulation of the origin of the solar system by Dole. The figures at the left indicate random numbers determining the injection of condensation nuclei into a condensing solar nebula. [From S. Dole, *Icarus* 13(1970): 500.]

that makes us think planetary systems might occur there. We realize it is not possible to observe directly by optical or by radio means the existence of planets. The only technique we have at the present time is inferring the existence of an unseen companion around a star whose motion is very carefully observed. This is a very difficult technique that can be applied only to the very nearest stars. Only a very small sample of stars can be observed in that manner, and even then it has to be a rather massive unseen object that is responsible for the observed very small variations in the position of the star. The majority of such cases, as now seen, involve a companion whose mass is much greater than the mass of our planets, and so very likely does

not re...   ...our planetary system. There
is just   ...star, in which the mass which
deflec...   ...comparable with the mass in
our s...   ...de Kamp has investigated this
very   ...ram (Figure 5) of the deflec-
tions...   ...r from the mean path that it
shou...   ...ad no companion. Notice the
smallness of the deflection involved. It is observed by means of
3000 photographic plates, on which he has very carefully measured the position of the star. He then interprets this in terms
either of one rather eccentric companion of a mass somewhat
larger than that of Jupiter, or in terms of two more circular
orbits of two planets whose masses would be quite similar to
the masses of Jupiter and Saturn. But Barnard's star itself is far

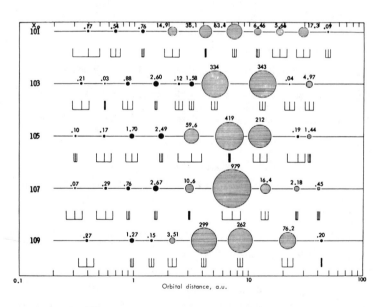

Figure 4. Another set of simulated solar systems, after Dole. [S. Dole, *Icarus* 13(1970): 501.]

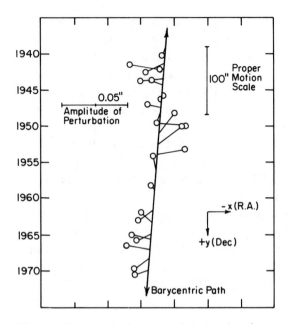

Figure 5. The perturbations, over three decades of measurement, of Barnard's star, due to one or more dark companions. The dark line shows the average motion of Barnard's star with respect to the background of more distant stars. The circles joined to this line show the perturbations from the mean motion due to the dark companion(s). After Peter van de Kamp.

less massive than the sun, about 15 percent of the mass of the sun and, therefore, these orbits are very much closer in, although approximately similar in period to the orbits of Jupiter and Saturn.

I think that there are many other interpretations possible for the types of companions that would account for these data and you mustn't take the Jupiter and Saturn interpretation too seriously. Of course, you must remember that the occurrence of bodies like Jupiter and Saturn in the vicinity of a star would not give any very strong clue, either, about whether bodies like the earth exist there. Nevertheless, if a concentration process begins it is very probable that the same situation would occur there as here, namely, that the refractory materials would condense near the star, and the lighter materials further away. So if large bodies like Jupiter and Saturn do in fact exist around Barnard's star closer to the star than the comparable planets are here, it is possible or probable that silicates and iron would have condensed also even closer to the star.

For our present discussion it is important, then, to realize that the circumstances that give rise to the condensation of silicates and iron are the very ones that would later lead to a biologically reasonable temperature regime. This suggests that if we are discussing a solar system in the vicinity of a much cooler star, planets like the earth would be appropriately closer to the star. It might not work out exactly, but it is in the right sense.

One final point concerns the spin of the planets. The inner planets of the solar system are very seriously affected by solar tides and have adopted states of rotation that are quite dominated by this effect. If we were interested in planetary systems much closer to a smaller star, then we would certainly have to worry that their spins would be much more dominated still by these tides.

It used to be thought that the sun would dominate in the sense of making the near planets rotate in such a way that the same hemisphere faced the sun all the time. That would make the planet very hot on one side and very cold on the other, a somewhat hostile planet for biology. Luckily, we have discovered, in the case of our own near planets, Mercury and Venus, that this is not so. Mercury was caught as it slowed its spin because of the tides—it was caught in a particular resonance so that it turns 1.5 times per orbit, and we now understand that that is really a very probable situation in any close planet.

This type of resonance prevents a relaxation to the synchronous mode of rotation. There are many resonances other than the one that Mercury is in that could equally well have served to arrest its mode of rotation.

Venus is in a more complicated situation still, but it also is not in synchronous rotation with the sun. Therefore, it is my opinion that at the moment we do not have to worry that a planetary system much smaller than ours, one with planets much closer in to a smaller star, would all be in synchronous rotation.

This is where we stand. You are all equally free to make up your mind how probable it is, from this information, that solar systems occur commonly around other stars.

OLIVER: I have some further information that might be of interest on the possible planetary architecture of Barnard's star. The original analysis by Van de Kamp that was published in 1963 assumed one planetary companion; the solution indicated an eccentricity of 0.65 for this one body. A later publication by Van de Kamp in 1969 assumed two bodies with zero eccentricity. Both of these solutions yielded about the same error residuals.

A more recent solution which has not yet been published, but which has been submitted for publication in *Icarus,* made by Suffolk and Black,* assumes three planets and yields residuals which are substantially smaller than either of the two previous solutions. The three planets seem to exhibit ratios of distances which are 1.55 for the first two and 1.61 for the second two. Both of these are similar to Bode's Law for the solar system. The innermost planet in this system is at a distance from the star of 1.8 astronomical units; the outermost at 4.5. Thus, there seems to be some indication that the planetary system may be scaled in proportion to the dimensions of the star itself. The

---

* This paper has since been published: D. C. Black and G. C. J. Suffolk, Concerning the Planetary System of Barnard's Star. *Icarus* 19 (1973): 353–373.

masses are 1.26 Jovian masses for the first, 0.63 for the second, and 0.89 for the third.

MOROZ: What does Professor Gold think about the origin of the moon?

GOLD: I believe that the satellites must all have originated in the form of clouds of small particles that came to be on certain planetary orbits and that gradually accumulated into a few satellites. The movement of small particles in the solar system is influenced both by gas drag and by the Poynting-Robertson effect, the dynamical effect of sunlight, and it is conceivable that small particles, instead of crossing the orbit of a planet and moving inward toward the sun (as is generally thought), would instead accumulate around the planets in part and subsequently form single bodies. It is not possible to think of the formation of the satellites in the same way as the formation of the planets around the sun, because it is not possible to think that the planets contained sheets of gas around them sufficiently dense to accumulate into satellites. The gas would not survive. You cannot think of a Jupiter system of satellites having been formed in the same way as the planetary system was formed around the sun.

TER-MIKAELIAN: You gave us very interesting calculations. Did you in these calculations take into account the loss of hydrogen? Can you propose any explanation for the loss of hydrogen?

GOLD: We know that the hydrogen must have been there if the solar composition of material was responsible for the formation of the planets. The present Uranus and Neptune are made from something which is very much richer in carbon, oxygen, and nitrogen than in hydrogen and helium. So the hydrogen and helium must have been lost from that region of the solar system. The amount that is lost can be judged from the amount of carbon, nitrogen, and oxygen that is present in the solar composition, and the amount is very large.

The present sun would not have been able to supply enough energy out there to evaporate all this hydrogen from the solar system. It is, therefore, a problem to know where the energy came from that took all this hydrogen away. It could be either that the sun at an earlier epoch was sufficiently brighter, or it could be that hydrodynamics in the Galaxy caused the gas in the outer part of the solar system to be swept away.

CRICK: The question is whether you took it into account in your calculations.

GOLD: Dole's calculations did not correctly include the hydrogen loss but merely had a turning off of the total amount of gas available.

GINDILIS: Will stars of earlier spectral types have adequate time scales for evolution on their planets?

GOLD: My view is that we should, in the first place, discuss time scales that are comparable to the one on Earth. Of course, it is entirely reasonable to discuss shorter time scales and expect in some other circumstances that perhaps life might evolve very much faster. On the other hand, the total number of massive stars is very small so that, if you are considering the more massive stars and a shorter time scale, then you will consider a very much smaller class of possible habitations for life than if you go in the other direction and discuss the less massive stars, of which there are many, many more, and whose time scale is very long.

MARX: Would you comment a bit on the rotation of Venus?

GOLD: The rotation of Venus is a most remarkable phenomenon. I didn't mention it in detail because, being so remarkable, I think it is not of very general validity in the discussion of other planetary systems, and the rotation of Mercury is not of that kind. The rotation of Mercury is of the kind that I would imagine would be very common in other planetary systems.

Venus appears to have started its life with a retrograde spin, to have been slowed down by very intense tidal friction with the sun, and then it appears to have been captured by resonance with the earth. The small tidal field of the earth in the position of Venus seems to have been enough to lock Venus in such a way that on successive closest approaches to the earth it faces the earth the same way around each time. There are many such modes, of course. It happens to be captured in the fifth such mode, but it is so close at the present time to this precise condition, better than one part in 2000 in rotational velocity, that we must suppose that it has, in fact, been so captured.

We have discussed in some detail how such a small effect as the earth's gravitational field could be important. At this point it appears that it can only be important if the solar tidal effect was almost zero because of the opposite influences of the solid body tide and the atmospheric tide. For the earth, we know that the atmospheric solar tide is in the sense of speeding up the spin of the earth with energy supplied by the solar heating in phase with the gravitational effect. That is in the opposite sense to the type of friction for the solid earth which tends, of course, to slow down the spin. For the case of the earth, the atmospheric tide is less than the solid body tide by a factor of 2. For Venus, it is quite possible that the two tides are comparable and also in opposite senses; if that were to be the case we could understand that the small additional effect of the earth's tide would at particular resonances dominate and capture Venus to that spin. That seems to have been the case. It is a very complicated and very particular circumstance.

GINZBURG: What is the rotation period?

GOLD: Two hundred forty-three Earth days.

AMBARTSUMIAN: Since Professor Gold discussed only the theoretically possible schemes of development of our sun and solar-type dwarf stars, it may be desirable to discuss briefly the results obtained in the course of investigations over the past few decades, based on a very large body of observational data.

The theoretical schemes can be elegant or they can be consistent, but unless they are based on a body of factual data they cannot be considered as relating to reality; whereas observational findings do always relate to real stars. I am speaking of the real stage in the development of stars in the lower part of the main sequence, that is, stars that have a mass of the order of our sun.

It is evident that the earliest stages of the development of such stars, the so-called T Tauri stage, directly follows the formation of a star. This early stage is marked by an extreme instability, a lack of steady state, and a pronounced change in the luminosity of the star in both the optical and the ultraviolet range of wavelengths. During this stage around the star there arises a sometimes fairly dense and even opaque gas cloud or shell, which subsequently disperses. Probably there is an intense ejection of particles, hydrogen, etc., in this stage.

This stage, which is probably very uncomfortable for life, is followed by the flare star stage. Apparently most stars, if not all, pass through this flare stage. At any rate, we have no evidence to the contrary, but we do have proof that the majority of stars pass through this stage. These recurrent flares are very pronounced at first. Then they gradually subside until they tail off to the flares that we observe in the case of our own sun. For nonastronomers, I will add that these flares last about half an hour to an hour; the ascending part, the rise, is very quick, a minute or two, and then there is a gradual decrease in the luminosity of the star. The energy of the flares is, however, very great, and sometimes the increase is one hundredfold or more compared to the initial luminosity of the star.

The energy of the flares is extremely large. It is of the order of 10^{36} or 10^{37} ergs. The flares gradually decline with the lifetime of the star, that is within a period of many millions or billions of years.

It appears that it is possible to say, although the body of evidence so far is insufficient, how the energy depends on the mass and age of the star. I will write this formula now, but I would like to stress the fact that this is a very rough approximation and a more accurate relationship will require the accumulation of further evidence:

$$E_{\max} = CT\mathfrak{M}^{-K}. \tag{2}$$

Here $E_{\max}$ is the energy of the flare, and I am speaking here of the maximum energy, because along with maximum flares there can be less energetic flares of that same star during its development; $\mathfrak{M}$ is the mass of the star, and T is its age; K, the index here, is somewhere in the vicinity of 1 but most probably between 1 and 1.5; C is a constant of proportionality.

This formula may not apply for the entire range of possible lifetimes of the star but it does apply in the interval between 1 million years and several hundred million. The intensity of the flares does decline, I repeat, very gradually, but some stars, such as the Hyades, for example, still possess a very high intensity of such activity.

If we are considering planetary systems around such stars, we must naturally bear these flares in mind. It is up to the biologists to decide whether these flares have a completely lethal effect, wiping out all life, or whether, conversely, these flares may have a stimulating effect leading to reactions that may give rise to life.

Equation (2) applies to the short wave part of the spectrum, the blue, the violet, the ultraviolet. We cannot give an exact account of the other part of the spectrum, but we assume that such flares are also accompanied, as in the case of our sun, by the emission of high-energy particles, and this must also be borne in mind. If you take into account that in the case of young stars this is all increased by a large factor, a factor of several thousands, then you are able to draw conclusions about the intensity of the phenomenon in that case.

So, I wish to emphasize that the development of young stars is marked by such flares and that the intensity of these flares decreases gradually.

I want also to say that recent papers by the British Astronomer Royal Wooley and several others indicate that there is a very large number of such young dwarf stars around us. The percentage is about one-third or one-fourth. One-fourth of the dwarfs surrounding us are about 10^8 years old, which is very young. How are we to reconcile this? After all, the age of the Galaxy runs to billions of years. Therefore, we are forced to conclude that star formation was not a uniform process and that in recent years, during the past 100 to 200 million years, this process was particularly pronounced. I think this is very interesting because possibly there were other periods of such intense star formation. It is possible that the sun is a very interesting star in the Galaxy.

I would not like to begin a controversy here about the origins of planetary formation, but I would like to add the remark that I do not particularly favor the condensation theory which was mentioned here by Professor Gold and several others. I would like to say that there are also, in my opinion, alternative explanations of this phenomenon. I would like to see all the findings of astrophysical observations and research reflected in our views of the origins of the formation of planets.

MORRISON: Is it possible that the energy of flares does not vary as a rather high power of the stellar mass as the luminosity does, and, therefore, that the relative intensity of flares grows as you go to less and less luminous stars?

AMBARTSUMIAN: I guess that the relative amplitude grows but this does not mean that stars with less mass have bigger flares.

MORRISON: But I think it does mean that their near environments are less steady for the cooking of these nice chemicals for the origin of life. Whether that is good or bad, I don't know.

GOLD: I am very glad that Professor Ambartsumian gave the astronomical part of the discussion, which really should have come ahead of my discussion of the condensation of the planets. I think that I should integrate this discussion a little into mine by a few remarks.

First, of course, one does not suppose that in the T Tauri stage, in the extremely violent phase of the early star, there would have been any suitable situation for forming the planets. I suppose also that the extremely violent flare situation would be a disturbance to the origin of life because of the very hard radiations that are contained therein; but remember I mentioned the problems of how a cloud of gas came into existence around the star before it was given the angular momentum that took it far enough out from the star to make eventually the major planets. We all think that this process is connected with the T Tauri phase.

We would then like to understand from the astronomical discussion also whether it is only for stars below a certain mass that a suitable cloud is generated which can then later be transported out to the far distance of the solar system. I hope that the astronomical researches, such as the ones Professor Ambartsumian mentioned, would in the course of time give us an indication of where the break point is and whether it is indeed in the same location as the break point of the stellar velocities that I described.

You remember the graph (Figure 1) of mass against angular

momentum. We would very much like to know whether this break point near spectral type A5 coincides with a difference in behavior in the T Tauri phase.

The last point Professor Ambartsumian made was that he does not favor the condensation theory of the planets. I would merely say that the only thing that I tried to indicate was that we were pretty sure that the terrestrial planets were formed out of solid particles and that the giants were then formed out of ices that later accumulated gases onto them. I know of no alternative discussions at the present time but if Professor Ambartsumian has some to offer I would certainly be glad to hear them.

AMBARTSUMIAN: The flare stage lasts for something like 100 times longer than the T Tauri stage. We observe, and can observe for the present, only flares in stars of low luminosity, for reasons explained by Professor Morrison—because the normal luminosity increases in proportion to a rather high power of the mass. For this reason we cannot observe flares in stars of F or G type even if they are brighter than those of stars of K spectral type.

GOLD: The flares you cannot see, I know, but the extent of the T Tauri phase you might still be able to recognize.

AMBARTSUMIAN: T Tauri phenomena are not observed in the case of high luminosity stars.

GOLD: Does the nonobservation stop at this mass?

AMBARTSUMIAN: Around F spectral type.

GOLD: That is pretty close.

AMBARTSUMIAN: As I said, I did not want to start a controversy on the question of the origins of planetary systems. I merely indicated that I was slightly skeptical of the condensation theory.

The body of evidence that has accumulated in recent years appears to indicate that the explosions and other phenomena that we observe in the nuclei of galaxies may be connected with planetary origins, and this may indeed be a more widespread phenomenon in nature.

HUBEL: Probably all the astronomers know the alternatives to condensation, but as a nonastronomer I don't have any idea what the list would be. I would like to hear a possible list of alternatives to condensation.

AMBARTSUMIAN: The disintegration of superdense bodies. The condition could arise that we begin from the formation of the nucleus of the galaxy, and then to stellar and planetary formation, a process taking place in the reverse direction. That is what I was implying. There are several articles on the subject. I have no real theory on planetary formation, but there are several such theories which we could consider.

MORRISON: Very likely this indicates that even in those well-determined portions of equation (1), there are still many things to be learned. I hope also that some people will begin to see why I never feel at home except on a dG star.

Another point was made in passing by Professor Ambartsumian, which seemed to single out the last few hundred million years in our stellar neighborhood, which again suggests to me that there are phenomena disturbing stars connected with galactic rotation, presumably the spiral arm rotation; it is very nice to live on a pleasant little G star in the disk. As far as we know, we are only accidently in a spiral arm.

MARX: Life started on Earth some 4 billion years ago. May I ask, how did the sun look 4 billion years ago?

MORRISON: Is one of our astronomical colleagues prepared to give an account of the early sun, and just how flarelike it was?

DYSON: I am not an expert, but the brightness was something like 80 percent of what it is now and, as far as I know, the flare activity would not have been enormously greater than now.

SAGAN: The more recent models of the evolution of the sun show an increase in luminosity of something like 30 to 40 percent from 4.5×10^9 years ago to the present time. If one were to believe these solar evolutionary models, and assume that the earth's albedo, surface infrared emissivity, and atmospheric composition were constant with time, then solar evolution implies that the global temperature of the earth was below the freezing point of seawater some 2 billion years ago.

There is a range of ways out of this evident contradiction, a contradiction because there is excellent paleontological and geological evidence for extensive liquid water as long as 4 billion years ago. Changes in the albedo work the wrong way; at lower temperatures, higher albedo; and one has very little room to maneuver the composition of the earth's atmosphere. CO_2 is strongly buffered; and at lower temperatures there would be less water vapor, not more.

The solution requires some atmospheric constituent which, if present in small quantities, can fill the large infrared window at 8 to 13 microns at the Wien peak of the Planck emission from Earth.

There is a gas which in small quantities does this absorption, namely ammonia. Something like 10^{-5} of ammonia in the early atmosphere solves this problem very nicely.

What is pleasant about this is that the following consequences follow: First of all, it is evidence for reducing conditions in the early history of the earth which are very suitable for the origin of life; second, it means that the temperature of the earth was somewhere around the freezing point of water at the time of the origin of life, which is very useful for the preservation of synthesized organics; third, it means that extensive oxygen in the earth's atmosphere was not present until 1 or 2 billion years ago at the very earliest. This is an example of how astronomical evolutionary events can be strongly coupled to biological evolutionary events.* I propose that there are other such correlations in our subject.

* Further discussion of this problem has been published: C. Sagan and G. Mullen, Earth and Mars: Evolution of Atmospheres and Surface Temperatures. *Science* 177 (1972): 52–56.

MAROV: I would like to add a few words to what has been said by Professor Sagan. The assumption that the sun 4 billion years ago had an energy density 30 percent less, roughly, is also brought out by various calculations on the Venus atmosphere. According to these calculations for radiative transfer of solar radiation for different mixtures of gases in a prevailing CO_2 atmosphere on Venus, the increase of solar luminosity over geological time was just enough to convert a rather earthlike Venus to the existing conditions on that planet, through a runaway greenhouse effect.

MOROZ: I would like to speak briefly about the possibility or impossibility of observing planets in other solar systems. What are the facilities that we need for detecting the emission of a planet from the background of the stars? The importance of this problem is obvious. Astronomers are aware of the situation that has arisen, and the paradox is that we know more about the origin of stars today than we know about the origin of our solar system. We are able to observe stars at different phases of their evolution, whereas the planetary system, the solar system, is before us at one particular moment only. For this reason, everything that Professor Gold said here today, he could have said ten years ago. Planetary cosmogony is marking time.

What is needed, therefore, is the observation of planets in different planetary systems. This is a fantastically difficult problem, but nevertheless I will venture to offer some relevant observations. Other technical solutions may also be possible but I will discuss only one.

Let us assume that we have an interferometer of the type used by Michelson for his measurements, a stellar interferometer. What should be the dimensions of such an interferometer and what would be the wavelengths for such work?

We can visualize this work as follows. Consider a star as a very bright source, with planets in movement around it. We are observing it by means of a stellar interferometer. There are two mirrors and a system for directing the emission into one telescope. If we observe the direction toward the star very accurately—a very difficult problem—the flux from the star will be maintained stably. In its movement, the planet will intersect the lobes of the interferometer. We will have a very slight fluctuation of the interferogram on the background of a very intense emission of the star.

So, what are the characteristics required for such an apparatus? Let us assume that the star has exactly the characteristics of our sun and the planets the characteristics of our earth. In that case, the ratio of the flux from the planet to the flux from the star would be 10^{-9}. That is a very substantial difference. We have a very faint signal superposed on the background of the star's emission. What number of quanta do we need to detect the signal? The minimum number of quanta we have to gather to detect the signal, determined by

$$N = (\Delta n/n)^{-2}, \tag{3}$$

is 10^{18}.

In order to have a signal to background noise ratio of about 3, we need 10^{19} quanta. What we have from the star, if it is at a distance of 10 parsecs, is 3×10^5 quanta per square centimeter per second. Here I am considering the complete emission of the star and not just the visible spectrum.

Evidently we must have a product of the area of the telescope by the observation time equal to the ratio of these two quantities. This is 3×10^{13} square centimeter · seconds. We cannot increase the accumulation time here very substantially because an earthlike planet will be moving with finite speed and the maximum time would be of the order of one hour, 3×10^3 seconds. From this we get an area equal to 10^{10} square centimeters.

In other words, we need an optical telescope of diameter ~ 1

kilometer, and this is not only impossible today, but I would say impossible for the next thousand years. Nevertheless, we can move into a different part of the spectrum where the ratio of the flux from the planet to the flux from the star is more favorable. The distribution of energy in the spectra of the planet and of the star differ substantially. The net spectrum will have two peaks, one in the visible corresponding to the star, the other in the infrared, corresponding to the planet. For the earth, that second maximum would be somewhere in the neighborhood of 10 microns.

This substantially alters the picture, for in this case the area-time product is 3×10^{10} square centimeter $\cdot$ seconds, and we will be satisfied with an area of 3×10^6 square centimeters. This would mean a diameter of the order of 20 meters and I think that such infrared telescopes will appear in our lifetime, in the foreseeable future, for these wavelengths.

I appreciate the fact that there are very many other technical difficulties involved. First of all, the base of the interferometer has to be sufficiently large. It would run to tens of kilometers, but I think that such a system could materialize somewhere, on the surface of the moon, let's say. In any case, such a system would have to operate in outer space, for the simple reason that if we were to base such telescopes on the earth, the sensitivity of the receivers, even if they were ideal receivers, would be limited by the effect of the earth's atmosphere, its background effect. We would need receivers with a threshold sensitivity of 10^{-15} watts (hertz)$^{-1/2}$. This is two orders of magnitude better than the characteristics of the best present infrared receivers, but I see no fundamental objections here, provided such a system is projected into outer space. I think in the foreseeable future it would be quite feasible.

KELLERMANN: What about longer wavelengths? How would they affect the situation?

MOROZ: The ratio would not become worse; it would remain roughly the same. But the energy situation would become much worse; we would be losing a substantial portion of the planet's energy. It would decrease as λ^{-3}. If we proceed, say, from 10 microns to 1 millimeter, we would lose a factor of millions in energy. I think that would be fatal.

SHKLOVSKY: If this hypothetical instrument were functioning, how many stars would we be able to observe?

MOROZ: All the stars in a 10 parsec radius—I don't remember how many that is.

SHKLOVSKY: Two hundred.

BRAUDE: Have you made any calculations for nonthermal emission?

MOROZ: No, I have made no such calculations but probably nonthermal emission could be used as well.

GOLD: Have you thought of the use of a Michelson interferometer for astrometric purposes? It is conceivable that we could measure the number of interference fringes between a star that we want to look at and a neighboring star, and thereby have an enormously more accurate way of doing astrometry than we have at the present time.

MOROZ: Undoubtedly.

MORRISON: Very closely related to these proposals, it seems to me that there is the bare possibility to improve astrometry, if we are lucky, by using very long baseline interferometry if the M dwarfs flare occasionally as some are known to do. We already have excellent interferometry with very long base lines but only in the 10-centimeter wavelength range. Very likely steady emission of either planet or star in this frequency band is negligible. But over the course of years, even quite steady stars do show flare-ups. Since a very long base line device already exists with a resolution of 10^{-4} seconds of arc, it may be, if we are lucky, that we will be able to work on Barnard's star. We are thinking about it.

MOROZ: Small stars, yes, but I am convinced it won't apply to planets. The flux is very faint.

BURKE: Doctor Moroz, it was my impression that you assumed that all of the signal from the star was received by the detector in the interferometer. Actually, for nearby planetary systems, the angular separation of a major planet and the star would be several tenths of a second of arc, so the analogue of a solar coronagraph technique could certainly be applied; the diameter of the needed telescope diminishes directly as the signal to noise ratio improves, so you go from 1 kilometer to 1 meter. I don't think you have to wait a thousand years for that.

MOROZ: You may be right. I didn't consider this possibility. I did consider that the entire response would be received and I thought it would be difficult to separate the flux of the star from that of the planet even at one-tenth of a second. We will always have instrumental scattering. Possibly a coronagraph will give us a tenfold gain; possibly a gain of two orders of magnitude. I think that in the infrared we will have the same gain, thanks to the spectral range.

PARIISKY: It seems to me quite obvious that if it should prove possible to build a large Michelson interferometer in the 10-micron wavelength range, we would either see nothing—at any rate in our galaxy—or we would see planets only, because the stars themselves can be resolved, the baseline being sufficiently large. The contrast can be increased by our selection of the range, and by increasing the resolving power of our instrument. An example in this case is the work already accomplished with large baseline instruments for discrete sources near the sun, experiments for checking the general theory of relativity. However, this method of direct detection of planets is clearly limited in sensitivity.

Another method has very considerable advantages in that it is practically independent of sensitivity. If it were possible to realize a very large obstacle interferometer, it is easy to see that the limiting factor would be the diameter of the planet rather than the distance to it—the diameter of the star rather than its brightness. I refer to astrometric measurements by means of a Michelson-type interferometer, employing nearby stars as reference sources.

I have the following estimates of the possibility of such interferometric measurements, in terms of the limiting detectable planetary mass. For very bright stars, say O stars, the limiting mass is equal to that of Jupiter for distances up to that of the Andromeda nebula. For G stars, the limiting mass is about one-fifteenth of Jupiter's mass. For white dwarfs, it is of the order of the mass of the earth. And for neutron stars of a small diameter, where great accuracy would be needed, it is of the order of the mass of the moon.

This method thus appears to compete, at any rate, if not to be more promising than the others proposed. Evidently it would be expedient to follow several avenues of research.

Finally, I would like to say that in my opinion large optical interferometers should be considered not only in space or on the moon, but also on the earth's surface. The October 1970 issue of the journal *Science* has a paper on a Michelson-type interferometer, operated in the optical range in the United States, with a 1-kilometer baseline and phase preservation for the light wave. As to the problem of the influence of the atmosphere, this influence can be substantially reduced by using reference points or by other methods already under discussion in the literature.

MINSKY: Couldn't one use a planet like Pluto as a very long-range linear coronagraph for interferometry? If one could get a telescope at the opposite side of the solar system and maintain it (since solar gravity is so weak there) by an ion rocket, one could get a very good signal-to-noise ratio.

SAGAN: There is another, much less expensive, proposal for detecting extrasolar planetary systems, made by Professor Frank

Rosenblatt of Cornell who has just died in a tragic accident. This paper has been published in *Icarus* this year.* The abstract reads:

The transit of a star by a dark companion or planet results in a characteristic colorimetric "signature" which should be detectable by photometric means. This signature, which is due to differential limb darkening in red and blue light, takes the form of a slight shift toward the blue as the planet crosses the limb of the star, followed by an abnormal reddening during the transit of the center of the star, and finally another shift toward the blue as the planet approaches the far limb. An analysis of this signature, the probabilities of detecting such transits, and possible instrumentation for this purpose are discussed. A system designed to yield one or more planetary detections per year is believed to be feasible at moderate cost. Such a system would consist of three wide field telescopes, at well-separated sites slaved to a central computer.

SHKLOVSKY: I fear that this project is handicapped by several defects. The activity of the star, sunspots, will cause color changes. It is enough for this to be of the order of 10^{-15} and the whole thing would be botched.

SAGAN: Rosenblatt discussed that, and many other sources of noise. Once a suspected detection were made, one would know what the period of the hypothetical planet was and then one could look for the transit the next time around. The proposal is well worth serious attention.

If I may, I would like to introduce at this time the question of G star chauvinism. Some of us have assumed for quite some time that the stars to be looked at are the stars of essentially solar spectral type. Stars of earlier spectral types have been excluded, as Professor Gold said, on grounds of age—that is, not enough time for life to evolve on their planets. But I think we have not been as careful as we ought to be about considering stars of later

* F. Rosenblatt, A Two-Color Photometric Method for Detection of Extra-Solar Planetary Systems. *Icarus* 14 (1971): 71–93.

spectral type. The fundamental reason that stars of later spectral type are of interest is twofold: First, most of the stars in the sky are of late spectral type and, second, those are the oldest stars. So if we believe that there are critical time scales—for example, if the origin of intelligence takes a few billion, but perhaps 10 billion years, then the most likely place to look for intelligence is on planets of stars of late spectral type. That essentially means M stars.

It was thought, at least by some, that many stars could be excluded because they were too cold—M dwarfs—but what is remarkable from the reduction of the Barnard's star perturbation data is that in all solutions the planets of Jovian mass are much closer to the parent star than in our own solar system. Even if this were not the case, an M0 star provides an equilibrium temperature at the distance of Mercury which is about that of the equilibrium temperature of the planet Mars in our solar system; and there has been some speculation about the possibility of life on Mars.

If, in addition, the semimajor axes of the planetary orbits are correspondingly shrunken, if they huddle in close to the parent star, then there may be several planets of each M star which are suitable for biology. If this is the case I wonder if we should not concentrate our attention, at least to some degree, on M stars. I would hope, for example, that Barnard's star might be a prime target in any search mode that was initiated.

SHKLOVSKY: We must bear in mind that G and F type stars comprise about 20 percent of the stars. That is, there is a colossal number of such stars, and even consideration of the F and G stars alone would therefore not limit our problem at all, for practical purposes. Although of course it would be very interesting to examine dwarf systems, too.

SAGAN: Unless there were a critical time constant. Unless we are early. Is it possible that intelligence has arisen in our system in a statistically unlikely early moment? If on the average one needs 8 or 9 billion instead of 5 billion years, then we must go to stars of appropriate age.

AMBARTSUMIAN: Recently there was a paper by Wooley and collaborators which contained weighty arguments in support of the youth of M stars, especially those that flare up. It is maintained that they have ages of 10^8 years or so—those observed in the vicinity of our sun, at any rate.

SAGAN: Does that apply to Barnard's star?

AMBARTSUMIAN: It is a statistical calculation.

SLYSH: There are very cold stars also, and I think we need not at this point consider the difficult problem of whether they have planets. We can also consider the problem of the existence of life and intelligence on the stars themselves. All the conditions there are as suitable as on the planets.

KARDASHEV: I would like to support what has been said here about the need to give attention to cold stars. It seems to me that, generally speaking, cosmology should consider the question of the formation of planetary systems without stars. In principle, of course, condensation can take place, condensation of low mass bodies, and they must survive. Internal radioactivity gives a stable steady course of energy that could support the development of life. Questions about radioactive sources will be considered later. At this stage I would like merely to say that the possibility of the existence of some planets without stars appears quite feasible and deserves consideration.

I would also like to remind you that many radio astronomers today consider that detecting a planetary system is much easier at the moment of planetary formation. There are suggestions that at the moment of planetary formation, the system may emit intense radiospectral lines such as the water vapor line or the interstellar hydroxyl line. In that case, the energy at the infrared range mentioned by Doctor Moroz may be converted into a very narrow frequency range and this would substantially improve the signal-to-noise ratio. Therefore, studies by an interferometer of anomalous maser sources may in the near future yield very interesting information about planetary formation, too.

I would also like to add that at the moment of the formation of planetary systems there may also arise very voluminous electrical charges, this because at the moment of formation, the electrical conductivity of a cloud of gas and dust would be very low; the formation of large spatial charges would give rise to a very strong nonthermal emission, which might be a detectable signature of planetary formation.

MINSKY: On the question of life on starless planets and cold stars, a source of energy is obviously required, but also a source of free energy, some sort of differential, and it would be hard for sophisticated chemical reactions to occur without something like a source of photons that represent some energy at a much higher level than that on the planetary surface. So radioactivity internal to a planet wouldn't do much good and again on the surface of a cold star thermal equilibrium would not help.

MOROZ: A few words about the possibility of the existence of isolated planets in interstellar space and about the possibility of life on such planets. The fact is that to support a temperature of $300°K$ on the earth about 10^6 ergs per square centimeter per second comes from the sun. The internal flux or flux from inside the earth is about 100 ergs per square centimeter per second, and to maintain the same temperature in interstellar space we have to increase the radioactive flux by 4 orders of magnitude.

What does that mean? If the thermal conductivity of the rocks is the same as on the earth, and on the earth we have a gradient (if my memory does not fail me) of about ten degrees per kilometer, there we will have a gradient of some 1000° per kilometer for the same 300°K. In other words such a planet would not survive in the solid state. At any rate, such a body would be radically different from the earth and it seems extremely doubtful whether it could support life of the type with which we are familiar.

EXTRATERRESTRIAL LIFE

$$n_e, f_1$$

SAGAN: If we are going to make some even tentative judgments on the likelihood of life elsewhere, we must decide to what extent the life we are familiar with is characteristic of all possible life. It may be that the kind of life that exists on the earth is only one small subset of a vast array of possible biologies, or it may be that life everywhere has to be in some sense the same kind as on Earth.

The history of this problem is of some interest. For example, there is a famous book published about 1912 by Lawrence J. Henderson, called *The Fitness of the Environment,* in which Henderson concludes that life necessarily must be based on carbon and water, and have its higher forms metabolizing free oxygen. I personally find this conclusion suspect, if only because Lawrence Henderson was made of carbon and water and metabolized free oxygen. Henderson had a vested interest. Can we make some sort of objective judgment free from chauvinism, independent of our prejudices? I merely want to raise, not pretend to be able to solve, this problem.

Here are some of the parameters involved in trying to approach this problem. The first question is the phase of the interaction medium in which the chemistry of the organism takes place. Many people would say that the solid phase prevents sufficient reaction rates to be useful for biology unless it is on the verge of a transition to the liquid state. On the other hand, gas phase media are probably unsuitable because the interaction products are not maintained. This then leaves the liquid phase which, oddly enough, is what we ourselves use. Is this liquid chauvinism?

An alternative possibility is a plasma medium, as popularized by Fred Hoyle in his science fiction book, *The Black Cloud*. The principal difficulty with this fully ionized organism is that there is no way for it to arise. Indeed, it seems to be immortal, and tied to the steady-state cosmology. I don't propose to examine further questions of the biological prospects in a plasma environment, but if the answer were yes, that would then open up a wide range of astrophysical environments which we now consider to be closed to biology.

In the parochial approach to the problem, we would demand that a planet have an appropriate liquid, possibly of high dielectric constant but not necessarily, and also with a large stable liquid range. If we also demand that this liquid have high cosmic abundance, then water is the only possible candidate material. But have we covered all cases, or merely succumbed to water chauvinism?

Now we come on to the question of the free energy exchange required for living organisms. This question was raised very interestingly earlier in our discussion and in part answered by Doctor Minsky. I would like to do a semiquantitative approach. A rough estimate of the maximum theoretical efficiency of a heat engine is something like $\eta = (1 - T_2/T_1)$, where T_1 and T_2 are the temperatures of the source and the sink, respectively. For

example, on Earth terrestrial photosynthesis is driven by a
6000°K blackbody, the sun, and the plants are immersed in a
300°K environment, the earth, so the maximum possible efficiency is something like 95 percent. In fact, the actual efficiency
of plant photosynthesis is somewhat less for other reasons.

Let us now consider a hypothetical plant on a hypothetical
planet. You can have plants on the earth without animals but at
the present time not the other way around, so it is appropriate to
consider plants. They are the fundamental energy reservoir for biology on Earth. Now consider a planet independent of a star and
let's imagine, as Doctor Moroz proposed, that its thermal conductivity is astonishingly high or, alternatively, that it has many
more radioactive heat sources than the earth, so it is at a temperature of 300°K. If we imagine 300°K to be the sink, there is,
of course, no source; but alternatively, we could imagine that the
300°K temperature is the source and that the planet radiates to
the 3°K blackbody background of space. Then we find η could
be as large as 99 percent. I have no idea how likely it is that
organisms can cool themselves off to the 3°K blackbody background, but my intuition makes me somewhat skeptical.

A related possibility is life on cold stars. There is no solid surface and the appropriate temperature range is somewhere in the
atmosphere of the supposed star, where the opacity is large. Accordingly, the hypothetical organism can drive its heat engine off
two temperatures which are about an optical depth apart. But
from the anticipated opacities one optical depth corresponds to
a very small temperature difference, and so I conclude that life
on such a cold star, as originally proposed by Harlow Shapley,
is not a very tenable hypothesis. But is this planetary chauvinism?

The next item on the list of properties relevant for biology is
atomic constitution. Here, the great variety of compounds employed in terrestrial biology is due to the bonding of carbon.

There are frequent remarks in the literature that no other atoms provide an adequate complexity of molecules, but because of a kind of carbon chauvinism, many alternative chemical systems have not been adequately examined. The chemists have understandably been interested in examining their own chemistry, and large ranges of possible other molecules have not been sufficiently investigated.

A related point has to do with the rate of chemical reactions. One finds in the literature statements on the Q_{10} of biological reactions, a measure of by what factor the rate of chemical reaction changes with a $10°C$ change in temperature. From reading the literature one would think this was a constant of nature. In fact, it represents a kind of observational selection, room temperature chauvinism. Those chemical reactions which go at reasonable rates at low temperature have already gone to completion at room temperature; those chemical reactions which go at reasonable rates at high temperature have negligible rates at room temperature. Chemists on planets at other temperatures will draw very different conclusions about the Q_{10} of chemical reactions. There is a wide range of activation energies of chemical reactions, and I think it would be a great mistake to think that necessarily the temperatures that prevail on the earth are a definite requirement for life elsewhere. A related point which I will make a little later is that the temperature environment of a given planet is a function of both location and time.

If we are going to talk about life elsewhere, we are surely considering a system of some formidable complexity, and a standard problem, both in the theological and in the biochemical literature, is how systems of such complexity could possibly have come into being. Let me give a numerical example. A simple protein might consist of 100 constituent units called amino acids, of which there are 20 biological varieties. Therefore, the probability of the chance assembly in the appropriate order of such a molecule consisting of 100 amino acids is 20^{-100}, or something like 10^{-130}. Well, it is clear that one could randomly assemble all the elementary particles in the universe a billion times a second for the age of the universe and never get this protein. Vastly less probable yet is a given human being. The molecules which determine the heredity of a given human being are the nucleic acids. Their constituent units, about which Professor Crick will shortly tell us a little more, are the nucleotides. A single human chromosome contains about 4×10^9 such nucleotide pairs. There are four possible nucleotide pairs. Therefore, a rough estimate of the genetic unlikelihood of a given human being is $4^{-4 \times 10^9}$ or roughly $10^{-2 \times 10^9}$. How is it possible that molecules of such breathtaking implausibility could ever have come into being? The answer was provided more than one hundred years ago by Charles Darwin. The preferential replication, the preferential reproduction of organisms, through the natural selection of small mutations, acts as a kind of probability sieve, a probability selector. It is only through enormous numbers of deaths and an enormous amount of time that we have reached the complexity we are at. There is no good theory that can predict how long it takes to get a certain complexity, but the complexity of the biological molecules and the genetic material is so large that it implies a very long period of natural selection. There is no doubt about the fact of evolution, but there are still sizable questions on the mechanics of the evolutionary process.

One final point: There are striking commonalities among the organisms on the earth. Not only are they all carbon based and aqueous but they all use the same molecules for transmission of genetic information, they all use the same molecules for molecular catalysis. In addition, the code book which transcribes the genetic information into the enzyme catalysts is the same, so far as we know, in all organisms on the earth. This is usually called the "universality" of the genetic code, from our point of view perhaps too large a term, but still a very remarkable fact. While there may be many organisms on other planets which are very similar to the organisms on Earth, at least in their fundamental biochemistry, we cannot exclude the possibility of extremely different kinds of organisms.

STENT: I would like to ask about dark stars. It is not clear to me why one couldn't have local temperature fluctuations on a dark star during which time the local temperatures would rise much higher than the surround, and during that time chemical compounds containing a high degree of free energy bonds could be synthesized, which would then serve as some kind of fossil fuel on which organisms could feed and draw as a source of free energy.

SAGAN: A small population might barely be maintained. A large population runs the risk of malnutrition. It is the same problem as was at one time considered in trying to postulate the existence of organisms in interiors of asteroids. Suppose there were a large quantity of organic compounds locked away in the interior of an asteroid, produced there in the early history of the solar system by the same kinds of prebiological organic chemistry I will describe shortly; why couldn't there be life inside that asteroid if the temperatures were suitable? There is no question that there could be life under those circumstances until the finite food supply was exhausted. What worries me about your suggestion is that the fluctuations in the availability of food produced on this star could wipe out the entire population.

GOLD: I feel obliged to go back and make a comment on the energy situation. I had merely said that it was convenient to have heat flow from the higher temperature source and go through a body which had a certain lower equilibrium temperature. That is what we have on Earth. But other possibilities undoubtedly exist. We can derive free energy from the surface of a body that has an internal heat source, be this radioactivity or be it energy production inside of a star. The only demand we would have is that the organism that does this job must not be immersed in a fluid of high opacity to the radiation appropriate to the temperature at the surface of the body.

On the earth we are immersed in air which is fairly opaque in the infrared. But if we had very little atmosphere, for example, it would be very easy to have a plant grow that exhibits the underneath of its leaf to the ground and the upper side of its leaf to space. That is the easiest way to derive energy from the plants, and the vegetation would grow spaced appropriately and covering as much of the ground as possible.

It is also possible to think of free energy in a body in which the organisms are immersed in an opaque medium (as was suggested by Doctor Sagan), but, for example, where there was a regular procession of temperatures. Suppose I sit immersed deeply in the atmosphere of a pulsating star; I can derive free energy by having a chemical change take place which is coupled with the equilibrium at each temperature extreme, and which can then have a time lag in reaching equilibrium. The problem that the food supply would run out would then disappear.

A similar remark applies for a planet which has a very large stellar energy flow falling upon it, higher than we have in the case of the sun, but which dramatically changes the equilibrium temperature of the atmosphere during the day.

I don't suppose there is life on Venus but at some depth in the Venus atmosphere we could have that situation, although very little high-energy photon flux is likely to be there. But there could be more favorable atmospheres from that point of view, so an opaque atmosphere is a perfectly possible biological habitat if it is subject to diurnal temperature variations. I express this because it seems to me we may have been too parochial in thinking that the particular type of energy that we use on Earth must be used everywhere else.

Incidentally, for those who thought that I was chauvinistic
in discussing only earthlike biological habitats earlier, let me
stress that I claim priority for having invented the notion of
Fred Hoyle's black cloud—but not, of course, that I contributed
to the excellence of his writing.

SAGAN: I may well stand guilty of the same chauvinism I was
inveighing against. As all our discussions will show, it is very
difficult to avoid arguing from terrestrial analogy. My only substantive response to what Professor Gold has said is that some
of the circumstances he has proposed, for example, pulsating
stars at $300°K$ surface temperature, are unlikely to be very
widely abundant.

CRICK: There is one remark I should like to make in amplifi-
cation of what Professor Sagan has said. This is the point that
the molecules that we have in biological systems on Earth have
a handedness. They are right-handed rather than left-handed.
Let me clarify the point. If you take a molecule of nucleic acid
and look at it in a mirror, you do not find that molecule on
Earth. So this is another thing, to add to what was said, which
is the same for all organisms on Earth. But I suggest that we
do not spend time on wondering why it is one hand rather than
the other.

I would like instead to discuss the biological system we have
on Earth, described in general terms, and in particular how we
should look at it when we consider the origin of this system. The
first requirement (which has been mentioned by Doctor Sagan)
that I would like to stress as a general property of any biological system is that it must have abundant versatility. In other
words, the system must be able to do a very great number of
very different things. It is of some interest to ask how the present system achieves this.

To discuss this, we have to ask what sort of activity we mean.
The activity which I think we are almost certain to require is

catalytic activity. This needs a three-dimensional structure because, although (as we have heard) it must take place in a liquid phase, we need some things which are like solids in a liquid phase, not only for catalytic activity but also in order to preserve the genetic information. The solution of this problem is that it is done by polymers; that is, proteins, and nucleic acids. The method that nature uses in building three-dimensional structures is not got only by using combinatorial methods (which was the point Doctor Sagan was making); there is an additional device.

It is not easy to copy a three-dimensional structure. The device used by life on Earth is to store the information in the nucleic acids, which have a one-dimensional structure which is particularly suitable for replication, but (in the form we have it) one which is not particularly versatile in catalytic or other activities; and then to translate from one language, the language of the nucleic acids, which has four letters, to the language of the proteins, which has twenty letters, to make a type of molecule which will fold itself into a three-dimensional structure. This is a general description of life as we have it on Earth. Now we can consider what characteristics of the system are essential.

I shall assume for the moment that we think only about natural selection and I will come back to this point. We have to ask, what are the properties we need to have a molecular system which shows natural selection? The first property we need is that we must have geometrical replication. It is not enough to have arithmetical replication, which is like the production of a newspaper from a printing press when you make many copies from one thing; here you must be able to make copies of the copies.

We now have to discuss mutations—that is, changes, accidental or otherwise, in the sequence we are talking about. It is a second requirement of our system that the replication mechanism should be able to replicate the mutations, that is, the "mistakes."

The next requirement is that whatever is produced, as we have seen from the earlier part of the discussion, must be able to influence the surroundings in a versatile way. There are other requirements I will not discuss, such as keeping one whole unit of the system in one bag, in one place, but I will leave that point on one side for the moment.

The proteins, which are very good at catalytic activity, do not appear, in the form in which we have them, to be very good as a simple replication mechanism. Consequently nature has used this device of two languages, one of which is good for replication and one of which is good for expression, and has devised an extremely complicated apparatus to translate from one language to the other, the results of which are our genetic code.

Now we must consider the problem of origins. To simplify ideas, we think of the moment in time (which may be longer than a moment) when natural selection first started to operate, although it may have been in a very primitive and inaccurate form. It is conventional for people considering the origin of life to say that when that point has been reached, the problem is solved. The reason is that anything that happens before that moment must arise by chance. However, one must remember that we live in a world which is structured by the nature of chemistry and physics; for example, the benzene ring is something which is stable because of the nature of chemistry. A very important part may have been played in the origin of life on Earth by the catalytic action of certain minerals which also exist because of the rules of chemistry. Consequently, we must not use the word "chance" too loosely.

Having said that, and having said that much of the earlier chemistry to make what I think is called the "soup" was available, we still do not see an easy path, with a reasonable degree of probability, which would lead us from the "soup" to the point at which natural selection would start. A rational man, having only the information we have at the present day, might reasonably conclude that the origin of life is a miracle, but this again reflects our ignorance on the subject. The point that I am making adds up to the following conclusion: It is not possible at the moment, with our knowledge of biochemistry, to make any reasonable estimate whatsoever of the factor f_l in equation (1). One feels a strong psychological prejudice that this factor will have a value near unity, but we have only one example before us, which is not enough on which to base any probability at all. We must conclude that until we have further information, we cannot really guess about the matter.

I have described the nature of the system on Earth. I will tell you my prejudice: I think it likely that, on another planet, life will also be based on small, complicated modules in a liquid, but I would not be prepared to discuss the details of the alternatives. What I have given you is the framework in which such a discussion can be conducted.

To finish, there are two points I would mention to show some unsolved general problems. The first question is the following: We see on Earth that there are two molecules, one of which is good for replication and one of which is good for action. Is it possible to devise a system in which one molecule does both jobs, or are there perhaps strong arguments, from systems analysis, which might suggest (if they exist) that to divide the job into two gives a great advantage? This is a question to which I do not know the answer.

The second question was one I mentioned at an earlier stage, that about natural selection. We could probably agree that the inheritance of acquired characteristics is at least not common on Earth, but a good general question is whether you could design a system which was based on the inheritance of acquired characteristics. This general question, so far as I know, has not been seriously considered. Similar questions have been answered; for example, R. A. Fisher showed that you could not have blending inheritance but must have particulate inheritance, so it is possible to produce general arguments in certain cases along these lines.

MINSKY: Biochemistry is advanced enough now that one could design a fairly complicated two-dimensional rather than three-dimensional string and substrate, so that it might be more interesting now perhaps to design very simple two-dimensional life forms, that is, look for the conditions of present organization, and once you have got it, work backwards and see if you can make a slightly simpler one. The path of working backward toward less complexity is probably easier than working up.

CRICK: I think one is agreed that one could think of a two-dimensional system, but you could generalize your remarks further. It is possible to imagine nucleic acids folding up in a three-dimensional way, and that could do what you describe.

MINSKY: But once folded, it takes extra mechanisms to replicate it.

CRICK: I think it is too easy to produce models and I will now produce one. Since we believe that planets rotate, you can go through all kinds of variations of temperature so that the same molecule at the lowest temperature could be in three dimensions and at high temperature in one dimension.

Can I put in a parenthesis? We have come to believe in the last few weeks that the control mechanisms in higher organisms depend on the intricate folding up of the DNA into three dimensions, but in this case with the help of one or two subsidiary molecules.

SLYSH: Doctor Sagan, in discussing black stars, you said that the amount of free energy on such stars is inadequate. I would like some quantitative estimates, if possible, as to what amount of free energy, say per unit mass or in some other way, is necessary for plants and organisms for life.

SAGAN: That will fit nicely into my next remarks. I have been asked now to say some words about prebiological organic chemistry. The question is, how can the molecules, these few molecules which are responsible for terrestrial biology, have first come into being?

The number of molecules actually used in biological systems is remarkably smaller than the number of possible organic molecules. There are billions of possible organic compounds. Less than 1500 of these are employed on Earth, and these 1500 are based upon 50 simpler building blocks. Of these building blocks, the most important are the amino acids, the building blocks of proteins, and the sugars and bases, the building blocks of nucleic acids. How can we understand the prebiological production of such molecules?

Suppose we were to imagine that they are made from the present environment; we could take a mixture of the gases in the present atmosphere of the earth, supply them with energy, perhaps an electrical discharge or ultraviolet light, and see what molecules we make. Under these conditions we make smog— ozone, oxides of nitrogen, and so on, and not what we are interested in making.

Suppose, then, we recall that the oxygen in the earth's atmosphere is produced by green plant photosynthesis; there could not have been green plants before the origin of life, so we now take the same gases without oxygen; that is, water, carbon dioxide, and nitrogen, and again supply that mixture with energy. At this point we make molecules like formaldehyde, not exactly where we want to go but on the way; because sugars are the polymers of aldehydes. The change that we have made is toward less oxidizing, more reducing conditions. Biochemicals have much more hydrogen, relatively, than the earth's atmosphere today.

At this point, we recall that the universe is composed primarily of hydrogen. The most abundant atoms in the universe are hydrogen, helium, carbon, nitrogen, oxygen, and neon. Since there is an excess of hydrogen, in cool bodies the molecules expected are the fully saturated hydrides of these atoms. Thus hydrogen will be present as the molecule (H_2), helium as helium, carbon as methane (CH_4), nitrogen as ammonia (NH_3), oxygen as water (OH_2), and neon as neon. It is, therefore, reasonable to assume that the early atmosphere of the earth had a composition of these molecules.

The theory of planetary exospheres clearly shows that hydrogen is able to escape from the earth, at present exosphere temperatures, in very substantial quantities over geological time, whereas atoms heavier than helium cannot escape at all. On the other hand, planets such as Jupiter have such large masses and such low exosphere temperatures that even hydrogen could not escape over geological time. It is therefore very comforting to find that the atmospheric composition of Jupiter is very likely just this mixture of H_2, He, NH_3, CH_4, Ne, and OH_2, but with water at such a depth that we do not see it spectroscopically because of its low vapor pressure at low temperatures.

With this encouragement we can then do the experiment one more time and mix together a mixture of methane, ammonia, and water, supply it with energy and see what molecules are constructed. The first such experiment was performed almost twenty years ago by Stanley Miller who found that amino acids were produced. Since then, a wide variety of such experiments have been performed, for example, in the laboratory of Professor Orgel and in our laboratory at Cornell, and what we find is that not only are amino acids made in high yield but also the nucleotide bases and sugars, and indeed all of the small fundamental building blocks of biochemistry.

To give one example of the yields of molecules which are implied in such experiments, we have done a set of experiments in which mixtures of these gases plus small amounts of H_2S, hydrogen sulfide, are irradiated with long wavelength ultraviolet light. H_2S is the photon acceptor. We produce amino acids with a quantum yield $\sim 10^{-5}$. We know what the ultraviolet photon flux of the primitive sun was from models of solar evolution, and so we can calculate how much amino acids were produced by solar ultraviolet radiation over, let us say, the first billion years of Earth history. If we were to assume, for purposes of calculation, that there was no destruction of amino acids (certainly not a correct assumption), then over the first billion years something like 200 kilograms of amino acids per square centimeter of the Earth's surface would have been synthesized. That is more than the amount of carbon available. If one puts in appropriate destruction rates of amino acids from the thermal degradation of such molecules, then one gets a yield which, if mixed in the present oceans of the earth, would give a few percent solution of amino acids.

Physics and chemistry are so constructed that very large quantities of the correct organic compounds—the ones that make us up—are produced under very general primitive planetary conditions. There is nothing in these experiments special to the earth, either in composition or in energy source. In fact, we have done experiments simulating the present atmosphere of Jupiter comprising the same gases under somewhat different conditions, and to no one's surprise we can still make lots of amino acids. Similar remarks apply to sugars and bases. In the early histories of planets throughout the Galaxy there must be efficient production of all these molecules. Precisely the molecules we need are made under the most general primitive planetary conditions.

This experimental result inclines some of us to think that the probability of the origin of life is rather high, although understanding the production of the building blocks of proteins and nucleic acids is not nearly the same as understanding the origin of life, and this latter fact I think was responsible for Professor Crick's skepticism about being able to establish any quantitative value for f_1 in equation (1).

The paleontological record has now been extended very far into the past; we now know of fossil microorganisms that are at least 3.2 billion years old. These are blue-green algae and bacteria, or at least have been identified as such by competent paleobotanists. These are very complex organisms, if one examines the microstructure and functions of the contemporary varieties. Also, the fossils that have been found to date are very likely not the oldest which will ever be found. Accordingly, the time between the origin of the first organisms, which must be much simpler than bacteria or algae, and the time of the origin of the earth, is not very long—only a few hundred million years or less. This, to me, speaks rather persuasively for a rapid origin of life on the primitive earth. Since we know of no special conditions on the primitive earth which could not be repeated on millions of other planets throughout the Galaxy, I have the sense, the feeling, that the origin of life is a very likely event. This is, of course, not a statistical probability in the sense of counting cases, but instead is a subjective probability in the sense described by Fine (Appendix A) which I mentioned earlier.

We already know that some kinds of meteorites have amino acids and large quantities of other organic compounds. We have good reason to think that comets have many organic com-

pounds. There is now a large and rapidly expanding volume
of evidence that a variety of organic compounds are produced
in interstellar space, including CO, HCN, CH_3CN, $HCHO$,
CH_3CHO, and HC_2CN, particularly in dense clouds. But all of
these pieces of evidence merely confirm that it is easy to make
organic chemicals, which we already know. What we need is
a planetary laboratory where the organic chemistry has been
left alone for a few billion years. That is what Mars, Jupiter,
and other planets provide, and if there is no inadvertent con-
tamination of such planets by microorganisms on space vehicles
from Earth, then it may be possible within the next ten or twenty
years to convert our estimate of f_1 from a subjective to a statistical probability.

Once a molecular system develops which is capable of evolution by natural selection, much of the subsequent development
of life is understandable, at least in principle. I consider the origin of self-replication and the genetic code, and not the origin
of cells, as central to the problem of the origin of life. Once
evolution starts, the selective advantages are apparent: for cell
membranes to protect living systems from the environment; for
the joining up of cells into the first eukaryotes and metazoa for
specialization of function; and for the origin of primitive neurons
for communication among cells.

MUKHIN: As Professor Sagan said, the commonly held view-
point today is that the predecessors of the more complicated
compounds were formed as a result of the action on a mixture
of gases of energy sources such as electrical discharge, ultraviolet radiation, radioactivity, shock waves and thermal radiation. I would like to point out that the possible correspondence
between laboratory experiments, laboratory models, and the
conditions that prevailed thousands of millions of years ago is
a question to which we do not as yet have an answer.

Also important is the question of the stability of the organic compounds that were formed. This is essentially evident from some of the proceedings at the Wakulla Springs Conference on the Origin of Life. There was a rather heated discussion there on the paper by Professor Fox.* The main objections put forward to the speaker concerned whether the laboratory conditions corresponded to the actual conditions many billions of years ago. I would like to offer you a model of the formation of simple organic compounds in the primordial ocean where the energy source and the source of the initial reactants were underwater volcanoes. I would underscore that in this model, the volcanoes are sources both of the initial material and of the energy necessary for organic synthesis.

Volcanoes are probably a sufficiently general factor in the evolution of the geochemical history of the earth, and we may consider this source to be responsible for the formation of organic compounds, organic molecules—as legitimate a source, in fact, as the ultraviolet radiation and other such sources. Within the sphere of action of such an underwater volcano, we have a sufficiently wide range of various factors. We may assume that the temperature ranges from 20°C to 1000°C, and if we regard the volcano as our reaction vessel, then the pressure there may be assumed to range from 5, say, to 500 or 1000 atmospheres. Finally, volcanoes are a source of gases such as ammonia, hydrogen, carbon monoxide, methane, hydrogen sulfide, the halogens, and many others. Water makes up 90 percent of the gases emitted by a volcano.

We thus have a sufficiently versatile mixture of chemicals and a sufficiently wide range of physical-chemical factors. The reaction may take place either in the gases or in the liquid phase, with the participation of solid state catalytic agents. Professor Crick mentioned the role of such solid phase catalysts in the

* S. W. Fox, ed., *The Origin of Prebiological Systems*. New York: Academic Press, 1965.

formation of organic compounds. Undoubtedly an eruption of a volcano is attended by the escape of a sufficiently wide range of minerals possessing such catalytic properties and a volcanic eruption releases a good amount of phosphorus, which we know plays an essential part in the life process.

More complicated compounds, such as the amino acids, the bases of the nucleic acids, and others can be easily produced from more primitive types of compounds such as formaldehyde, hydrogen cyanide, and others. Therefore, to establish the validity of such a model we have only to demonstrate the reactions that would lead, say, to the formation of HCN. Volcanoes produce ammonia and carbon monoxide, and we know that the reaction of ammonia and carbon monoxide inevitably produces HCN; likewise the reaction between carbon monoxide and hydrogen would inevitably produce compounds such as aldehydes. Professor Sagan and several others have mentioned in their papers that these compounds are the precursors of practically all the biologically active molecules.

Undoubtedly the model I am offering does not preclude the possibility of other mechanisms playing their part and possibly its only advantage over other models is the fact that it lends itself to experimental test, under natural conditions today.

ORGEL: Doctor Mukhin's remarks raise a very general question: Are our rather conservative views on the time and space required for the origin of life supported by any theoretical or experimental data? The usual assumption is that the evolution of life took place over a wide geographical area and took a period of hundreds of millions of years. I know of no experimental evidence pointing in that direction. I know of no theoretical computations which give any clue at all as to the time that the origins of life may have taken. I think it is an interesting challenge to ask why life could not have required one million years to arise in a small volcanic area or in some other small specialized area. Once photosynthesis had been invented, then life would have become independent of any of the special conditions which were necessary for life. I would like to hear the opinion of Doctor Sagan and others on this question.

SAGAN: First of all, there are *some* experimental data relevant to the question. We have a probable upper limit to the amount of time it takes for the origin of life, namely, something like a billion years, from the fossil record. That is certainly information of relevance. But the question you raise is, Might the time be much shorter than a billion years? I suggested that it may have been 100 million years, but maybe it was much less—six days was at one time a popular suggestion. Certainly it may have taken only one million years, but I don't see how we are going to know the answer.

On the question of special environments, that would be useful to consider if we were having difficulty understanding the production of these molecules under more general conditions. But we do not seem to have any such difficulty at least with the small building blocks. On the other hand, for polymerization, for example, it might be very useful to have special locales: Professor Crick mentioned one and alternatives have been mentioned by others—for example, on the surface of minerals like montmorillonite or hydroxyapatite. What I find especially nice about the latter suggestion is that it enables phosphorus to play a role in the origin of life out of proportion to its low cosmic abundance.

We have a tendency to think of planetary environments as homogeneous in space and time, but this is not correct. For example, the possibility of life on Jupiter is, in almost all astronomy textbooks, excluded by reference to very low temperatures, where these temperatures apply only to the outermost cloud layers. In fact, by the same argument life on Earth could also be excluded. The best contemporary models of Jupiter show that there is a very pleasant environment at some depth below the visible clouds where there is very likely liquid water, a temperature of 300°K, and just the array of molecules which we need for the origin of life, at least in the gas phase. That is an example of heterogeneity in space. Heterogeneity in time depends on the type of atmosphere involved. Through chemical reaction with the soil, by escape of hydrogen to space, and through the greenhouse effect, complex and interesting changes with time in both the surface temperatures and the atmospheric chemistry of planets are likely.

MORRISON: It seems to me that, though the discussion is very interesting, it does tend toward two distinct objectives without perhaps quite separating them. If I put it that a special micro-environment may be very helpful, the probability of the occurrence of life seems to me still to be raised by a general background of manufacture and synthesis of organic compounds by large-scale energy sources. For example, the solar input is, I suppose, a million times the energy input from volcanoes. That by no means shows that volcanoes or dry pools or geysers or seafoam or anything you have in mind might not be a very useful site. But since we have no evidence against using a long time, 10^9 years, it seems to me we can use the general background and hope that the appropriate conditions for the origin of life occurred perhaps in many ways.

So, I come back to the solar free energy, such an available energy source and still the primary biological energy source, though by no means the only one—there are microorganisms that work under ferrous-ferric conversion. It seems to me that there is still a strong case for saying that the presence of a star's radiation is a very important energy source on which other events may have to be superimposed for the origin of life.

GOLD: I would think that the question of volcanoes or other energy sources should be taken very seriously because of the extreme difficulty that early biology had. I would think that volcanoes, which are an obvious source of energy, both through the chemicals that they supply, which are not in equilibrium with the normal chemistry of their surroundings, and through the heat that they supply, might well be very significant sources of energy.

In addition to that, one can think of chemical sources of energy that derive from the different equilibrium conditions on different parts of the earth—for example, I can imagine that the chemical equilibrium high up on a river is different in some sense from that further down the river, and if this river transports some of the chemicals downstream, the availability of these chemicals that are locally not in equilibrium would be an energy source. These energy sources exist nowadays on the earth but they are rather small compared with the supply of photosynthetically derived energy. But in the old days those may well have been the important sources. So to Doctor Mukhin's volcanic story I would add consideration of the transportation of chemicals from one region to another where they are, being in disequilibrium, an effective energy source. Rivers flowing from a hot region to a cooler region would be a case in point.

ORGEL: I would like to take up another topic which I think is of great interest, namely, whether the factor f_1 is up the river. I know Doctor Crick wants to talk about it; also Doctor Sagan.

CRICK: I think we should underline the theoretical and experimental difficulties in explaining the origin of life on Earth. You have heard, I think, that there is really no serious problem about two things. One is the availability or the synthesis of the simple building blocks. The other is the supply of energy. We have not worried about that because we think there was enough chemical energy lying around and once the system had got started it could look after itself. The problem we have to worry about is how to get the system started, and that is a problem which could loosely be described as one of information rather than energy.

By "information" I mean joining together the building blocks into polymers and, secondly, into the right polymers. The difficulty we have is to produce polymers which can then go on and do some sort of replication. Polypeptides can be produced very easily but they cannot replicate. Again, the polymerization of nucleic acids, which can replicate easily, is not so difficult to imagine. Having made random polymers, we have to wonder very much about the probability of making polymers which either have the correct sequence or a correct family of sequences. It is this probability that we cannot make any reasonable estimate of at the present time—but I hope not forever.

I want now to take issue with Professor Sagan's position. Because we see at a period of 3000 million years back organisms already formed, and because we only have another 1000 million years before we can start making the soup, his argument is that the time to get to the takeoff point may have been only, as he said, 100 million years, although, as Doctor Orgel has said, it could have been even shorter. Therefore, the force of this argument is that even if it took ten times longer on another planet, it would not have made a lot of difference. But the factor of 10 is the one we are worrying about, and it would have made a lot of difference if that factor of 10 was a factor of 10^6.

What the present biochemical evidence shows is that there was in some sense a unique event. We know this because of the homogeneity of the biochemistry of different organisms, but we do not know whether that homogeneity was at the takeoff point. All we can say is that at some stage our ancestors must have gone through a small population, but we cannot be clear what that stage was; nor can we get an estimate of whether the unique event might have happened several times, because once you got to the takeoff point, this elementary form of life could spread over the earth and that might happen in a short time, shorter than the time for the event to recur. With this argument alone we can make no estimate of the frequency of the event. We have subsidiary arguments, for example that there could have been competition and one system would have won out. But all this does not help at all.

And now in order to display the difference between my position and Professor Sagan's, I have to make an analogy and I am sorry it is so conventional. Consider a man who has been dealt a hand of playing cards. The character of his hand is that he has to have one particular sequence, one particular combination of cards. We know that this is a rare event and it is not reasonable to try to estimate the probability of the event simply because it has happened to us. Professor Sagan's argument is that there are plenty of playing cards. But we only have a unique event and strict probability theory says that we are not allowed to deduce probability in that way. That is what is called statistical probability. We must therefore turn to his other concept, subjective probability, which is so much used by businessmen in the United States and scientists in the Soviet Union in designing their researches. But that subjective probability is based on the capacity of the human mind to see relationships between things, out of past experience, which are not clearly formalized, and this is the tool that all imaginative scientists use in the prosecution of their work. But in our present problem we do not have the experience on which to form these hunches. Therefore, in my view it is not legitimate to appeal to subjective probability and we find ourselves, therefore, I say, in a state of ignorance which we hope the future will reduce.

SAGAN: Professor Crick's provocative remarks relate not just to the factor f_l but also to the factors f_i and f_e and L, in equation (1), which are even more difficult problems, and which we will have to face in the following discussion.

I do not agree that we have no relevant information. In addition to the paleontological information which we have already discussed, there is work on the abiological replication of polynucleotides. We think we see a direct sequence from the primitive environment to crudely self-replicating nucleic acids. But the origin of the genetic code remains unsolved. It is certainly true that there are no experiments as yet on the origin from the primitive gases and waters of the early earth of self-replicating mutating systems which strongly interact with their environment. Professor Crick and I are playing different card games. In his interesting playing card analogy, I do not believe there is only one sequence of cards which wins. My expectation is that there are many paths to the origin of life, and that the joint probability that one of them has been taken on a suitable planet over billions of years is rather high. But how can we settle this difference of opinion?

The main consequence of this argument is, I believe, to make even more important the search for life on planets like Mars and Jupiter. In the case of Mars, there is a wide range of possibilities, from remnants of prebiological organic chemicals to contemporary microorganisms to the possibility of even more advanced life forms. Philip Morrison has stressed that the discovery of life on Mars would convert the origin of life from a miracle to a statistic.

CRICK: I want to add something because I think it has a general application to all the factors in equation (1).

What we are trying to estimate in that equation is the magnitude of the various factors, but there is another thing which we have to ask and that is the reliability of each estimate. What I was trying to say about f_1 was that the reliability of our estimate was very low. But that should not be taken to imply that the probability itself is very low. It is important to have these two concepts clearly in mind—to have a spectrum of possibilities, to each of which you can attach a reliability estimate.

SAGAN: With that I entirely agree. For completeness, we should at least mention the panspermia hypothesis. This is the idea that microorganisms may have been wafted from a planet of one star to a planet of another star, perhaps by stellar radiation pressure or perhaps embedded in an interstellar comet or meteorite. In discussion at lunch with Doctors Crick and Orgel I discovered that there is, at least in a minor way, an advantage in explaining the "universality" of the genetic code if one were to imagine the origin of life on Earth initiated by such a microorganism from elsewhere.

The panspermia hypothesis was originally proposed by Svante Arrhenius at a time when there was no information at all about how the origin of life could have occurred. It was a means of avoiding rather than facing the problem; namely, by proposing that life came here from somewhere else and ignoring any questions about how it arose in that somewhere else.

We have just completed a set of calculations using the full Mie theory, which gives the following result: Those microorganisms which are ejected by radiation pressure from a given solar system accumulate a stellar radiation dose both in ultraviolet and in x rays, which is a thousand to ten thousand times the mean lethal dose of the most radiation-resistant terrestrial organisms known. Those organisms which are sufficiently large not to be killed are not ejected by radiation pressure. Special organisms can be "designed" to avoid these difficulties, but they correspond to no known organism. Thus the prognosis for the classical panspermia hypothesis seems to be negative.

There are possibilities of organisms embedded in interstellar comets, as I mentioned, but the likelihood of such a transport is very small, and the accumulated radiation dose from cosmic rays and natural radioactivity is embarrassingly high.

THE EVOLUTION OF INTELLIGENCE

f_i

HUBEL: I think I should warn you at the outset that this presentation may not be very clear for two reasons. One is that I am still eight hours out of phase and I usually do not lecture very well at three in the morning; and, secondly, I had the fortune or misfortune to have breakfast with Academician Ambartsumian who insisted we have an Armenian breakfast complete with cognac. You will have to judge whether these two events are advantageous or disadvantageous.

A third difficulty is the problem with which I am confronted. In earlier discussions we got as far as the development of life, and now I am called upon to begin a discussion of the nervous system. But of course there is a large gap, because we have not even begun to discuss the evolution of a simple cell, to say nothing of a multicellular organism.

The problem that we wish mainly to address is the difficulty or ease with which an animal, such as a human being, can be developed. All of these problems, from the single cell to the chimpanzee and man, I must now briefly discuss. The strategy that

I propose to use is to say a few words about very simple animals and simple nervous systems, and then proceed immediately with a kind of caricature discussion of a complicated nervous system. I hope to spend most of the time trying to give an impression of what a nervous system in a higher organism is like.

One question which I will not deal with at all is the problem of learning and memory, about which I know very little. However, there is much that can be said, and this whole area will be discussed by Professor Stent.

Now to proceed: The simplest organism that we can talk about is the single cell which must fend for itself and solve all its own problems. It has to be very generalized and it has to be able to do many things.

Single cell animals, many of them, are able to move. They may have cilia or hairs of some kind which they wiggle and they may move in various other squirming fashions. Obviously, they are also able to respond to the environment. Many kinds of single cell animals—probably all—respond to changes in the chemical environment. They respond if you poke them with a needle, and many are able to respond, for example, to light.

At the next step, the multicellular organisms, these various abilities of animals tend to be partitioned into a number of different systems. For example, the organism will have an entire muscular system devoted to the problem of movement, and a number of different sensor systems that act as transducers in transmitting information from the environment. If the animal is of any degree of complexity, it will want to make some rather sophisticated movements. For example, even a worm will have circular muscles and longitudinal muscles so that it can shorten or lengthen or twist, or even make complicated movements such as occur in swimming.

At the other end, the sensor system also must soon develop degrees of sophistication in order to extract from the environ-

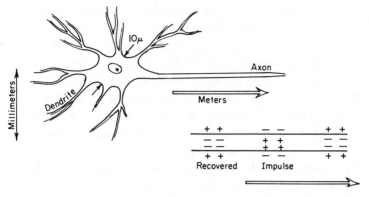

Figure 6. Schematic diagram of a typical neuronal cell with characteristic dimensions indicated. At lower right is a schematic representation of the propagation of an electrical wave along an axon.

ment information which is relevant to the life of the animal, so it is not surprising, then, that a special system must soon in evolution be devised to take care of sophisticated movements on the one hand and to extract information from the environment on the other hand.

Finally, the various systems of the animal—the digestive, endocrine, and so forth—must be correlated. All of these things, then, are what led to the development of some system which will handle information. For multicellular organisms, on this planet, the solution has been remarkably constant from one animal to the next. The main solution has been the nervous system, although one must remember there are others of importance, such as the endocrine system.

The next step, then, is to describe the properties of a single nerve cell and then to proceed to describe some simple properties of higher nervous systems. So let me begin presenting an example of a possible nerve cell, although there is an enormous variety of nerve cells. In Figure 6 is a drawing of a rather typ-

ical but perhaps exceptional nerve cell. It has many of the characteristics of cells in general. It has a nucleus and it has a cell
wall, but the cell wall is enormously increased in area compared
to that of most nerve cells with true life structures and rich
branch systems, which may, if you can see an entire cell, look
something like an oak or an elm tree. In fact, the comparison is
not bad because there may be just as many branches and processes.

In the cell body many of the processes that occur in most other
cells take place. But generally a rich number of branches and
processes come from the cell body, and these are usually devoted to the receiving of information. They are usually called
dendrites, and the endings of other nerve cells end in close
proximity to the dendrites, or to the cell body; most of the information comes to the dendrite or the cell body.

Coming from the cell body usually is a single process, long and
slender. This is called the axon. The diameter of a single body
will be about 10 microns. The territory occupied by the entire
branching pattern is measured in millimeters, but the axon may
be measured either in microns, or, in the case of an animal like
us or the giraffe, may be a number of meters long.

At its termination the axon usually breaks up into a number
of endings and these usually terminate on the cell body or on
the dendrites of some other nerve cell. [The connections are
called synapses.] The information comes in at the terminals, and
the sum of all the incoming information is integrated. Then
simple stereotyped signals are sent down the axon.

When the electrical signals reach the termination, usually a
chemical is released there. The chemical diffuses across through
the extracellular space to the next cell which must combine that
information with the information from perhaps many hundreds
of other incoming signals. Finally the decision is expressed by a
signal which goes along the axon.

The signal always goes in the same direction. The axon has a positive charge outside and a negative charge inside. During the passage of the wave signal down the axon, this charge is reversed. This wave of reversed signal proceeds down the axon at a rate of perhaps a meter per second. The signals are all determined by differences in chemical concentration of ions.

The nerve cells then can be looked upon as cells that are specialized to respond to chemical signals. The chemical, of course, is secreted by the preceding nerve cell and the response of the nerve cell is expressed in turn by the secretion of the chemical.

The one thing that I haven't said so far is that, depending on the chemical, the result of the secretion may be such as to make it more likely that the cell will fire; but the result may also be that the cell is less able to fire. Any given synapse will be either inhibitory or excitatory. Which it is depends upon the chemical that is secreted and on the properties of the membrane that receives the signal. A number of chemicals have now been identified and are known to be neural transmitters. We know, on the other hand, very much less about the chemistry of the full synaptic membrane, though this is under active investigation at the present time. The properties of the nerve that lead to excitation and inhibition, the changes in conductance of the membrane, the process of impulse transmission along the axon—all of these things are now quite well understood, thanks to the work of Hodgkin, Huxley, Katz, Eccles, and others.

Let us look at how these cells are put together in a typical organism. In Figure 7 I have drawn a number of specialized cells which we can call receptors. These are in many ways like nerve cells except that they respond to environmental information—light, mechanical deformation, etc. Like most nerve cells, they often have axons and the axons break up into a number of branches at their termination. These branches end up in other cells which are shown in very simple form, so you can see from

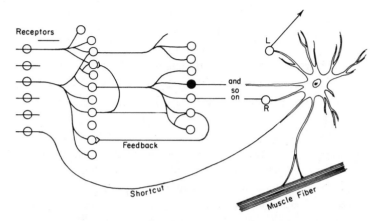

Figure 7. A schematic nerve net connecting receptor cells (for example in the optical system of the eye) with an output—in this case a muscle fiber. The cells marked L and R are responsible for distinguishing between motion in the visual field from left to right and motion from right to left. See text for further details.

such a diagram that one receptor can influence many nerve cells at the next stage and a given cell at the next stage then receives many inputs from many receptors.

The same process tends to occur stage by stage. Sooner or later one reaches a final stage—a nerve cell whose axon ends on a muscle fiber, so here you see the input and the output. Without the input, the animal is cut off from the environment and is nothing but a vegetable. Without the output it cannot have any influence on the environment and is also a vegetable.

So here we have the part of the nervous system that integrates the environmental information. Close to the end there are specializations for the organization of movement and somewhere in between one has everything else—memory, the soul, and perhaps even the social sciences.

We know a fair amount about what happens close to the input to the nervous system and we know in some animals a fair amount about what happens close to the output. But in between we know relatively little. However, the lesson that one can learn from looking at regions close to the input and close to the output may be relevant to what goes on in the middle.

You must keep in mind that this diagram (Figure 7) is a caricature. There are some exceptions. The axon may sometimes take a backward course with feedback or a parallel course. The circuit between the input and the output may be very short. The simplest example is the tendon jerk. A slightly more complicated one with a few intervening stages is the pupillary reflex when one shines a flashlight into the eye and the pupil constricts.

Now let me point out a relatively more complicated possible reflex to give you a feeling for what may happen at the input and output. At the output one is concerned with the organization of movement. For example, if you make a fist, obviously you do so by contracting many of the muscles in the anterior surface of your forearm. What is less obvious, perhaps, is that if that were all you did, you would also obviously have a flexion of the wrist joint.

Without giving it any thought, when you make a fist what you do automatically is to contract some other muscles on the posterior surface of the forearm and this helps, of course, to keep the wrist rigid. As anyone who has children knows, this is nothing that one learns, because when you stroke the palm of the hand of a newborn baby, the fingers close and a beautiful fist is made, and the result is not a flexion of the wrist joint. The machinery of this movement is built into us from birth.

Now a slightly more complicated movement is involved when we look at something with our eyes. The movements of each eye are controlled by six muscles. When we look to the left, two muscles contract and their antagonists relax. The two muscles that contract are not symmetrical muscles; they are different muscles, the lateral muscle in the left eye and the medial muscle in the right eye, and this is done with a great degree of precision and with no thought on our part.

Here one can imagine that a nerve cell that comes before the final nerve cell sends its output to selected groups of nerve cells which go to just the correct muscles. So such a cell as this, then, might be important for moving the eyes to the left and a different cell for eye movements to the right, and others for up, down, close or far. All of this is rather obvious.

Now let us turn to the input stage. You must remember that the inputs to any of these cells may be either excitatory or inhibitory. It came as a great surprise some years ago when people first put electrodes close to cells in the heart of the visual cortex that is concerned with vision. The most important thing that was found was that such cells, even though the path that leads to the receptors to those cells is very clear, gave no response to crude stimulation of the eyes. They were completely uninfluenced, for example, by shining a flashlight directly into the eyes of the animal. Of course, any cell here will get its input, will be influenced by many cells at the receptor level, some excitatory and some inhibitory, and these tend to balance each other in a very precise way.

The overall lesson is that general light levels are not very important to us. We all know that it is differences in contours falling on our retina which are the important thing in our vision, and to some extent the differences in wavelength. So one may have cells for the input which respond very brilliantly and fire like a machine gun to a green light, but their firing is stopped by a red light, and to a yellow light or a white light they give no response at all.

Other kinds of cells, and probably the most common kinds, respond to stimuli in some particular orientation in space. Some cells prefer one orientation, other cells prefer a different orientation. To handle all of the possible variables, such as orientation, wavelength, movement, and so on, obviously requires very many cells, but this need not disturb us because very many cells is precisely what our brains possess.

It is important to understand that there are not very many kinds of cells. On a wild guess, there might be three or four hundred kinds of cells, just as one could imagine three or four hundred different kinds of trees—oak, elm, and so forth—none of them precisely alike but any given kind easily recognizable.

To continue, then, with the reflex that I was building up, we can imagine a cell four or five stages into the nervous system which may respond when a spotlight moves across the retina from left to right but not from right to left. Any electrical engineer with a little imagination could easily devise a circuit to do such a thing, and some of the circuits in the nervous system are reasonably well known and rather simple.

If this cell responds only when something moves from left to right, we only need to connect it to the cell which is responsible for movement to the right and then we have a reflex in which something moving from left to right in a certain direction across the visual field induces a following movement of the eyes.

I have concentrated perhaps unduly on some of these simple things. Despite the fact that the qualities in our own brains that we are mostly concerned with at this conference have to do with speech or communication in general, or other intellectual functions, we know almost nothing about the neural organization of such qualities. Unfortunately, it is very difficult to do the required experiments in human beings, and perhaps we don't even know enough to ask the appropriate questions. Nevertheless, the lessons that can be learned from these simple things suggest that it will probably be possible in the next few decades to understand many more complicated things in terms of the simple building blocks of the nervous system—excitation, inhibition,

the nerve impulse, and the various circuits which I have just illustrated—many of which can be seen anatomically.

If we ask ourselves what possible alternatives there are to such
circuits, anyone's guess is of about equal value, provided he
knows enough to make guesses of this sort. It seems to me that
one thing that one must acquire is the ability to organize complicated movements. This is a problem in the distribution of information to various muscle groups. So we have to have many
information channels. Long before anything was known about
the nervous system people were already using wires to conduct
electricity and sending cars along roads and performing other
information theoretical tasks.

I suppose in a sense one can think of a beehive or a termite
colony where, in an abstract way, the individual insects would
be comparable to the individual nerve cells of our nervous system. It is really unnecessary to complete the thought. One can
be as abstract as one wishes.

For many classes of animals, it appears not to have been particularly advantageous to devise a nervous system that could go
far beyond the phase of simple reflexes, such as those that I
have illustrated here. A frog responds by touching or snapping
at an insect or, indeed, any black spot of appropriate size, provided that it is moving. One can easily imagine that nerve cells
in the frog's visual system do not respond to stationary black
spots, and this is, in fact, what we find.

Insects have been very successful, as a class, and have survived
for many millions of years without developing an intelligence
sufficient to manage radio telescopes. Compared to this, a chimpanzee is perhaps very close to what one would like to achieve.
A chimpanzee can do very complicated and imaginative things,
but I have not the slightest idea what it is that has caused one
group of animals to evolve along a direction leading to high in-

telligence and others to go the direction of the insect, ending in a sort of evolutionary cul de sac. The ideas that one hears about such things are not very satisfactory. It is said, for example, that the chimpanzee and certain classes of monkeys soon became very interested in swinging from one branch to another. This led to great facility in movement with their hands and in vision, but perhaps avoided specialization of other kinds, so avoiding these evolutionary dead ends.

I can't pretend to have any profound knowledge of this sort of biology, but I think these are the directions in which we need to be thinking when we try to imagine what problems there are in evolving a higher nervous system.

Of course, one big difference between a worm or an insect and ourselves, or a chimpanzee, is involved in the degree to which these different animals use anything that we would call learning, which Professor Stent will shortly discuss.

Finally, there is an enormous problem in any given animal of developing this nervous system from a single fertilized egg. How do all of these fibers know in which direction to grow once connections are established? This area is under very active investigation and we know many isolated facts, but not enough is yet known about nervous system development to help our deliberations here.

CRICK: Could you give us some idea of the number of nerve cells in, say, an insect and the number of nerve cells in, say, a man, or perhaps a part of a man? I would remind you that insects are necessarily small for other reasons.

HUBEL: Yes, this is very easy to do to order of magnitude. The number of nerve cells in an animal like a worm would be measured, I suppose, in the thousands. One very interesting thing is that we may point to a particular individual cell in a particular earthworm, and then identify the same cell, the corresponding cell in another earthworm of the same species. At the other extreme, for man there are probably no very good figures, but the one usually quoted is about 10^{10}. Take the human eye (Figure 8) and the retina as an example. The number of receptor cells in each retina is 125 million. There are many other nerve cells in the retina, probably millions more. Curiously, however, the number of optic nerve fibers which represent the output of the retina is roughly one million.

Now let us look at the cerebral cortex, which is a folded sheet of cells about 2 millimeters thick (Figure 8). If we consider the cortex as a plate (it is convoluted in order to get a certain area inside of our skulls), and if we look under one square millimeter of cortex, the number of cells is about 10^5. The amount of cortical surface area in our entire cortex, or at least in the cortex of Canadians, is about 1 or 2 square feet. This yields a total number of nerve cells in the cerebral cortex of around 10^{10}.

STENT: I have been asked to make some remarks about the

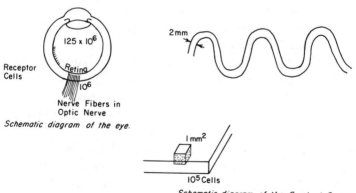

Schematic diagram of the eye.

Schematic diagram of the Cerebral Cortex

Figure 8. Schematic diagram of the numbers of nerve cells in the eye and in the cerebral cortex. There are 125 times more receptor cells in the retina than nerve fibers carrying information to the visual cortex from the retina. The nerve cells in the convoluted cerebral cortex are shown schematically at top right. Each square millimeter of cerebral cortex surface, if unfolded, contains below it 10^5 cells.

evolution of the nervous system, but it is very difficult to say very much that would be of help for the purposes of this conference. However a few things *can* be said. One is that the properties of the membrane of the nerve cell which allow it to generate electrical signals are of high evolutionary antiquity and antedate the appearance of nerve cells. For instance we encounter them already in such unicellular animals as Protozoa. The first multicellular animals of which we have knowledge, for example, jellyfish, already possessed nerve cells which have more or less the typical shape of the vertebrate neuron described by Doctor Hubel. These nerve cells were connected to form the simplest nervous system called a *nerve net*.

With the appearance of the earliest bilaterally symmetric animals, for example, the flatworms, a further advance of the nervous system took place; namely, the development of a *central nervous system*. The difference between the nerve net and the central nervous system is that in the latter there exists a higher degree of specialization of nerve cells, in that certain nerve cells bring signals only from the outside in, and other nerve cells bring signals only from the inside out.

Furthermore, invention of the central nervous system made possible the concentration of nerve cells in the anterior part of the animal, or the development of the brain. As in the course of this development the animal's nervous system and its behavioral repertoire becomes more complex, the number of nerve cells continuously increases from the approximately 10^3 nerve cells of a simple flatworm to the 10^{10} to 10^{11} nerve cells in man.

We may now consider a topic to which Doctor Hubel has already referred, namely, learning. Learning is the most complex manifestation of a more general attribute of the nervous system, namely, plasticity. Its plasticity allows the nervous system to have a history—that is to say, its present state depends on its past experience. It is obvious that plasticity, and in particular the capacity for learning, confers on the nervous system a great advantage from an evolutionary point of view.

But I want to make a point which is less obvious, namely that plasticity is not merely a fringe benefit but an essential ontogenetic feature of any complex nervous system. I will base my argument on some important experiments which Doctor Hubel, himself, has carried out. Contrary to his disclaimer, he is one of the very few people who do know something about learning. The only reason that I am presenting his experiment here and not he is that he does not like to philosophize.

Let us consider once more the visual system of the cat that Doctor Hubel has already described for you. Suppose the eyes of the cat fix on a light bar in the visual field. The light coming from this bar forms an image on the corresponding areas in the retinas of both eyes. The retina is a mosaic of photoreceptor cells, and light falling on the photoreceptor cells gives rise to electrical signals. The photoreceptor cells are connected through many intermediate stages to nerve cells in the cerebral cortex of the cat.

Each such cortical cell receives visual inputs from the retinas of *both* eyes. The retinal photoreceptor cells are connected to the cortex in such a manner that a given cell in the cortex gives rise to a train of nerve impulses when corresponding retinal areas in the two eyes simultaneously report the excitation caused by their illumination. For instance, when the eyes look at the light bar, a particular cortical cell gives an impulse train and the cat then knows that there is a bar of light at a particular position of its visual field and in a particular orientation. There are very many such binocular cells in the cortex, each of which speaks to one particular orientation at one particular position of a light bar in the visual field. But for the system to be workable, it is necessary that the two converging sets of photoreceptors which are connected to the same cortical cell do get their visual information from exactly the same position in the visual field. Because if the converging photoreceptors from one eye were seeing a position different from that seen by the receptors converging from the other eye, the cat would be very confused.

How can this system arise during the embryonic development of the cat? On the one hand, as Doctor Hubel has already pointed out, there must exist a process which precisely connects the appropriate sets of photoreceptor cells from corresponding parts of the two retinas to a given binocular cell in the cortex. Although we do not understand how this is done, we can imagine at least that the genes do it. But there is the additional problem of the physical optics of the eye, which themselves arise also thanks to ontogenetic processes directed by the genes. That is to say, the lenses of the two eyes must be exactly alike and be positioned precisely so that the image of a given point in the visual field falls exactly on those two corresponding points of the retinal receptor mosaics which the genes have managed to connect to the same cortical cell. However, for the genes to achieve such a high degree of perfect matching in the direction of two wholly independent ontogenetic processes seems well-nigh unimaginable. And hence it appears that in directing the connections between retina and cortex, the genes simply *over-connect* the system. That is to say, congenitally each cortical binocular cell receives inputs from much larger retinal areas in the two eyes than is actually compatible with sharp vision. This sloppy system is then refined by the earliest visual experience of the animal, by a process which defines corresponding retinal areas of the two eyes as those points in the receptor mosaics that, given the two lenses which the animal actually has, receive light from the same point in the visual field. That is to say, the animal *selects* among the surplus of existing connections just those which are actually workable and which actually bring to each binocular cortical cell a coherent visual input.

The work of Doctor Hubel and his colleague, Wiesel, has shown that during the first three months of the life of the cat there occurs a critical period during which the connections from retina to cortex are in a state of plasticity, and that only those connections survive the critical period which have brought a coherent input to a given cortical cell. This is a process that can be described as learning, in that the cat's cortex selects from an overabundance of connections those which experience teaches it are useful. Many other examples of plastic critical periods are known in animal development, although the case of binocular vision is the only one for which the neurophysiological basis has so far been established.

I have presented this example in such detail because I wanted to demonstrate why plasticity is likely to be a necessary attribute of the existence of any complicated nervous systems. In other words, the reason why learning is not just an additional, evolutionarily favorable fringe benefit is that without it the very genesis of complex nervous systems would be impossible.

To summarize, there is one hopeful aspect in estimating the factor f_i in equation (1). There is good reason to expect that if a complex nervous system develops at all, then it will also develop plasticity, which is a prerequisite for learning—which, in turn, is necessary for intelligence.*

MORRISON: Are any social insects known to exhibit plastic states?

STENT: Yes, in the case of bees, there is a form of learning. A scout bee can remember the direction and distance of a source of honey for some minutes, at least; it then returns to the hive, and does a dance for the other bees to indicate to them the spatial coordinates of the honey source.

* Editor's note: A later and related discussion which may be of interest to the reader is G. Horn, S. P. R. Rose, and P. P. G. Bateson, Experience and Plasticity in the Central Nervous System, *Science* 181 (1973), 506–514.

SAGAN: Don't flatworms learn?

STENT: That's a sensitive subject. It is operant learning. Training has been reported, of course, for many invertebrates. I think Morrison's question was not about plasticity in general but about the sensitive period. There, I don't know of any case in invertebrates. But, certainly, earthworms and snails can be trained.

SAGAN: Are there any cases of training in Protozoa?

STENT: Yes, they have been reported. It has been claimed that you can train Protozoa to swim up a glass capillary.

CRICK: Doctor Stent's point was that plasticity was a necessary condition. He did not say it was a sufficient condition, and we have not the time to discuss the complexity of the nervous organization which is needed to form concepts. Moreover, as Doctor Hubel said, this subject is in a very primitive condition and it is even scarcely rewarding for specialists. We see again that this is a field in which we need very much more research in order to have a proper understanding. However, it is a problem that is of such interest to all of us, and to many other people in the world, that it will be surprising if important progress is not made even in the more difficult parts of the subject within a period of ten or twenty years.

THE EVOLUTION OF
TECHNICAL CIVILIZATIONS

$$f_c$$

LEE: My task is to take the discussion a step further and describe the conditions which led to the origin of intelligent organisms capable at some later time of developing a technical civilization. I also wish to consider the question of how far we may generalize these conditions to life forms on other planetary systems. In this study we are limited to one empirical case, our own species, with perhaps the addition of some others such as the chimpanzees and dolphins, which are more intelligent than other animals. At first, this paucity of material is discouraging. But I am more hopeful, for we are armed with several powerful tools of analysis that can make our task considerably more productive of insights.

The first tool is the modern synthetic theory of evolution by natural selection.* The modern synthesis has so well accounted for the diversity of life forms on this planet that it offers us a

* See J. Huxley, *Evolution: the Modern Synthesis*. New York: Harper, 1942.

working beginning for understanding the nature of life forms on other planets.

The second tool is the theory and method of historical materialism.* This seeks to elucidate the general laws of human history and society. The method pioneered by Marx and Engels attempts to do for intelligent life on this planet what the Darwinian synthesis has done for life in general.

The third tool is the commitment shared by most of us to search for the broadest, most comprehensive generalizations that can be drawn from available facts. We believe that there are general laws operating in the universe and that our understanding of the universe advances as we reconcile new observations which are in apparent disagreement with the laws as currently formulated.

These three tools—evolutionism, historical materialism, and uniformitarianism—offer a basis for using our experience to shed light on intelligence as a general process. Living systems, both intelligent and nonintelligent, possess several properties that are relevant to our inquiry. First, every life form is adaptive and its adaptation is a product of its responses to changes in the environment over time. Second, every life form has a history of gradual change and divergence from an ancestral form. Every form, no matter how complex today, evolved from a simpler form. Third, a life form may exist for a very long period of time. And fourth, evolution, given long time periods, is a means for generating highly improbable results.

These four points offer a place to start our inquiry. For example, we can infer that any intelligent civilized form we encounter in space will have evolved to that state from a simpler nonintelligent form or will have been constructed or partly pro-

* See K. Marx, *A Contribution to the Critique of Political Economy* [1859]. New York: International, 1968; V. G. Childe, *Social Evolution.* London: Watts, 1951.

grammed by a being which itself evolved from a simpler form. I will discuss this latter possibility later.

Let us consider now the evolutionary history of our own species from a simple to a more complex form. Under what circumstances does intelligence evolve, and how did it arise in our species? Before we can answer these questions, we have to define what we mean by intelligence. Intelligence should best be regarded as an adaptation. Therefore I will discuss intelligence, not from a philosophical point of view, but, rather, as a zoologist would, who, when faced with an adaptation exhibited by an organism, has to specify what it is that this structure or behavior does for the organism. In this perspective, we can say that intelligence is the means by which an organism copes with a complex problem-solving situation. In a word, intelligence is an adaptation for more complex behavior. Intelligence is adaptive only in an organism which has complex behavior patterns involving many alternative choices. In other words, we can't conceive of an organism that can entertain five different courses of action when never more than two behavioral alternatives are open to it. Or to put this another way, a complex brain must be hooked up to a complex behavioral system.

In what areas of the life of an organism does one find room for the development of complex behavior? Not in such areas as locomotion, food-getting or escape from enemies. All these areas are extremely well programmed in the organism early in life and allow very little room for experimental departure. The main area that offers scope for complex behavior is the social field. It is only here that complex behavior can develop without endangering the survival chances of the organism. Therefore, we must look to a highly social group of animals for the rise of intelligence. The higher primates, including the monkeys and apes, are relatively large, long-lived mammals, high in the food chain, and these are the most intensely social animals in nature. Most monkeys and apes live their entire lives in strongly knit social groupings. These primates form firm and lasting attachments with dozens of other animals, and they interact with them hour to hour every day of their lives.* Also important for the development of intelligence is the prolonged period of dependency of the infant primate to its mother. This close bond offers the opportunity for the transmission of a great deal of learned behavior. In Doctor Stent's terms, we have a great prolongation of the plastic period. Thus, I believe that complex social life is a prerequisite for the evolution of intelligence, and this precondition should apply to intelligent life forms in the universe generally.

A second kind of prerequisite is physiological. The absolute size of the animal may prove important. Clever monkeys with stereoscopic vision and hand-eye coordination have existed for at least 25 million years. However, despite their cleverness, they were too small. They weighed a few kilograms and their brains were less than 100 cubic centimeters in volume. Since absolute size of the brain is important for the number of neural pathways, the actual emergence of human intelligence may have become possible only after our ancestors reached a larger size of about 30 kilograms body weight and 400 cubic centimeters minimal cranial capacity.

Also, the animal had to have an adequate blood supply to the brain (this disqualifies the giraffe) and, more important, the brain had to be placed in such a way that its expansion didn't interfere with the continued functioning of other systems, such as food-getting. The brain cases of most mammals are heavily constricted by massive jaw muscles and olfactory apparatus.

* See I. DeVore, ed., *Primate Behavior*. New York: Holt, Rinehart, 1965; A. Jolly, *The Evolution of Primate Behavior*. New York: Macmillan, 1972.

Thus, the early achievement of upright posture may have freed the brain for further development.*

These are the prerequisites, social and physiological; they are few in number. Other factors seem to me to play no necessary part in triggering the emergence of intelligence.

This process apparently occurred sometime before 4×10^6 years ago in a now extinct species of anthropoid primates of the family Hominidae that holds the current designation *Australopithecus*.

The timing of the appearance of man has been pushed steadily backward. Until recently the appearance of tool-making was thought to have occurred one million years ago. Then in 1959 the late Louis Leakey discovered *Zinjanthropus* in the Olduvai Gorge of the Republic of Tanzania. This was dated by the potassium-argon technique at 1.75×10^6 years. In 1969, in the Lake Rudolf area, Richard Leakey discovered hominid remains and stone tools in unquestionable association with a deposit dated at 2.61×10^6 years ago.† Hominid remains without stone tools have been found at the Omo beds in Ethiopia and dated at 3.75×10^6 years ago.

This creature, also known as the gracile Australopithecus, is known to us by the remains of over one hundred individuals from a dozen different localities in Africa. They were small, about 150 centimeters in height and weighing between 30 and 40 kilograms. They had erect posture, though it can be said that they could run better than they could walk. Their hand was like ours, able to grasp objects with a precision grip, and they made tools. But it is very curious that their brain was very small by human standards. The cranial capacity has been measured at

* See B. Campbell, *Human Evolution*. Chicago: Aldine, 1965.
† R. E. F. Leakey, Fauna and Artifacts from a new Plio-Pleistocene Locality near Lake Rudolf in Kenya. *Nature* 223 (1970): 223–224.

450 to 650 cubic centimeters. This size falls within the range of the living great apes.

This creature is the best candidate we have for the ancestor of the genus *Homo* of which we are the only surviving representative. So it is in the study of the behavior and adaptation of the gracile Australopithecines that we must seek the causes for the emergence of intelligent life on this planet.

What made it happen? I will offer three related explanations that are current in North American and British anthropology, in particular those drawn from the work of Professor Sherwood L. Washburn at the University of California and that of his students, of whom I am one. The first concerns tool-making, the second the transition to hunting as the mode of production, and the third the emergence of language.

The importance of tools in human evolution has long been known.* Tools were the first forms of extra-somatic energy harnessed by man. The invention of tools, in effect, moved the focus of the adaptive forces from the muzzle (face, teeth, and jaws) to the hands and brain.

We now know that man is not unique as a tool-maker among the higher primates. In studies of chimpanzees in the wild, Jane van Lawick-Goodall has found that chimpanzees regularly manufacture several kinds of tools. These include a stick for probing termite mounds and a sponge for drawing drinking water from inaccessible places. So the designation of man as *the* tool-making animal no longer strictly holds.† Nevertheless, the tool allowed man successfully to exploit a new ecological niche and thus had decisive consequences for human history.

---

* See F. Engels, *The Part Played by Labour in the Transition from Ape to Man* [1896]. Moscow: Progress Publishers, 1968; S. L. Washburn, Tools and Human Evolution. *Scientific American* 203 (1960): 63–75.
† See J. van Lawick-Goodall, Tool-using and Aimed Throwing in a Community of Free-living Chimpanzees. *Nature* 201 (1964): 1264–1266.

At a point early in the lower Pleistocene, man moved from a herbivorous to an omnivorous mode of subsistence with hunting as an important feature. The successful hunting of large mammals requires a level of social cooperation and coordination of movement that is unprecedented even for the social primates.

In this view, early man filled an ecological niche in Africa for a large omnivorous mammal living in the open plains, not in the forest. This adaptation combined the social solidarity and mutual protection enjoyed by the primate group with the large range, size, and high-quality diet of the large carnivores such as lions, hyenas, wild dogs, and wolves.

In order to fill this niche, however, early man had to develop methods of communication to a much higher degree than formerly. Not only did the hunters have to signal each other in the hunt, but there was also the question of distribution of meat to the nonhunting members of the social group: the females (who provided the equally important plant foods), the young, the old, and the sick who stayed behind. In effect it was the new division of social labor as well as the new mode of subsistence itself that generated the evolution of intelligence. Here we come to the heart of the matter: *Human intelligence reduced to its essentials is synonymous with human language.* Intelligence is improved communication, the transmission of more complex information from one individual to another.

In the first instance, language may have evolved in the calls exchanged by hunters in the hunt, but in the second instance, language became a means of mediating social relations within the group. This more elaborate signaling system had a feedback relation to the social system. Now a far more complex social life became possible and this placed an adaptive premium on a socially skillful animal as well as simply a better hunter.

The British ethologist, Michael Chance, has argued that *equili-*

bration, the ability to defer action until a more auspicious occasion, had a tremendous selective advantage in the evolution of man.* This is the ability to think one thing but to do another, or to think before acting, an ability that would be equally adaptive for women and men, unlike "hunting skill" which would seem to be selective only for males. Perhaps our colleagues in neurophysiology may be able to elaborate on the separation in the brain's circuitry of thought from action. Where in the early evolution of intelligence consciousness—that is, self-awareness— arises, is not possible to say. The late Irving Hallowell and others before him have argued that consciousness represents the true dawn of humanity, but how we are to pinpoint this event in the fossil record is an extremely difficult problem.†

One point is clear: Once language becomes established, it has its own logic of development. In fact, language becomes elaborated far beyond the adaptive needs of the organisms who possess it.

This problem has puzzled me. The !Kung bushmen of southern Africa, with whom I lived for several years, exhibit an intelligence entirely comparable to our own. However, they exist with a total material culture of less than one hundred named items! On the other hand, their communicating abilities are truly impressive. At night they sit around the campfire telling stories that are full of complex metaphor, humor, innuendo, and all the modes of expression that we have come to associate with the literature of an advanced civilization; yet these are people who practice no agriculture and have no domesticated animals except for the dog.

* M. R. A. Chance and A. P. Mead, Social Behavior and Primate Evolution. *Symposia of the Society for Experimental Biology* 7 (1953): 395–439.
† A. I. Hallowell, Self, Society and Culture in Phylogenetic Perspective. In Sol Tax, ed., *The Evolution of Man,* Chicago: University of Chicago Press, 1960.

Are we to conclude from this that the tremendous growth of human intelligence was largely for social and recreational purposes? The evolution of language and its relation to the adaptation of the early hominids is still very poorly understood. Much work remains to be done in this area. All one can say at this point is that language may have originated to make man a more perfect hunter but that the unanticipated consequences of this event were very great. Man's intelligence not only led eventually to the destruction of the hunting way of life that it was designed to perfect but it is now in danger of destroying the species itself.

However, while it lasted, the hunting and gathering adaptation was an extraordinarily successful one. For 99 percent of human history, man lived as a hunter. It was as a hunter that man underwent an explosive radiation that led him to occupy every corner of the habitable globe by 10,000 B.C., and, contrary to common opinion, man's numbers did not remain stationary during the Pleistocene period. It now appears that man's population increased one-hundredfold during the Pleistocene without, however, going over the carrying capacity of any specific area.

From the hunters we inherited several things: first, a pollution-free environment; second, an egalitarian social system; third, a good family life; fourth, a robust physique; and fifth, a taste for steak. Today we find ourselves faced with an ecological crisis and also the threat of nuclear annihilation, so the hunting way of life may yet prove to be our most stable and successful adaptation.*

How did we get from this pleasant state of affairs to the current predicament? It is involved with what happened at the end of the Pleistocene, 10,000 years ago, with the origin of agriculture and the rise of cities, states, empires and technical civilization,

* R. B. Lee and I. DeVore, *Man the Hunter*. Chicago: Aldine, 1968.

and this is an appropriate moment for me to turn the floor over to my colleague, Doctor Flannery.

FLANNERY: It is my job now to explain how man fell from this Garden of Eden, which Doctor Lee has described, into his present disastrous condition. I cannot estimate exactly what the probability of the rise of civilization is, because we do not know how many prehistoric cultures did *not* become civilizations. But my subjective feeling is that the chances must be quite good, because civilizations arose independently in many different parts of the world in prehistoric times; in Mesopotamia, India, China, Egypt, Mexico, and Peru. We do not know yet how independent all the cases were. We know that Mesopotamia, India, and Egypt were, in fact, closely related, but they were certainly independent from Mexico and Peru, where civilizations arose involving a completely different set of agricultural products, races, and languages.

In spite of the frequent rise of civilizations in prehistoric times, civilization is a very late event in human history. If we could compress all of man's life on Earth into one 24-hour period, civilizations have existed only during the last minute of that day. In Figure 9 is the time scale for the development of civilization in Mesopotamia as one example. It started at about 10,000 B.C. The dates on the timetable are only for Mesopotamia because the times are different for other civilizations.

If we start at 10,000 B.C., near the end of the Pleistocene, the first permanent villages appeared in the Near East by 8000 or 9000 B.C., before agriculture began. Agriculture had begun by 7500 B.C. Manufacturing of ceramics had begun by 6000 B.C., specialization of villages by craft activities by 5000 B.C. Social stratification followed very closely on this, metallurgy was developed between 3000 and 4000 B.C., and writing and bureaucratic civilization had appeared in the Near East by 3500 B.C. The total length of time involved was about 7000 years.

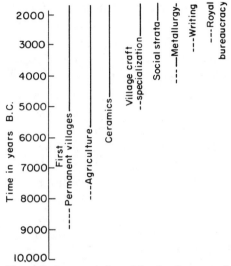

Figure 9. A presentation of the development of civilization in the ancient Near East. Similar sequences have occurred elsewhere but with different absolute time scales.

It is obvious that there was selective advantage for civilizations once they had arisen, because they continued to expand and to replace other ways of life, which is a process still going on. In much the same way that early life arising on the earth consumed the organic molecules out of which it had developed, the earlier civilizations expanded, consuming the products of their environment until they had essentially destroyed the environmental conditions in which they had arisen in the first place.

For the purposes of this symposium, probably the three most important questions are—

Why did civilizations not arise earlier in human history?

Why did they arise in the parts of the world in which they arose?

How and through what phases did they proceed?

The concept of the passage of mankind through stages of primitive hunting to primitive agriculture and the development of stratified society is something that has been considered by social philosophers for one hundred years. Over this same period, a sequence of events has been theoretically reconstructed by social philosophers, among them V. Gordon Childe, Lewis Henry Morgan, Friedrich Engels, Julian Steward, Leslie White, and more recently by Elman Service and Marshall Sahlins.

But these reconstructions are formed not on the data of prehistory but by looking at peoples in the world today who still live in ways that we imagine our prehistoric ancestors must have lived. In recent years, one of the roles of archaeology has been to test these reconstructions against the actual data of prehistory and to generate and test new hypotheses about the rules for the development of prehistoric societies.

What did all the prehistoric or ancient civilizations have in common? First of all, they all produced their own food by means of agriculture. They all had a hierarchical society with a professional ruling class. They all had very high population densities, compared with man in the Pleistocene. Most of them had cities or urban centers and state-supported arts, crafts, and architecture, and almost all had writing systems of various kinds, although these were very different from one region to another. The writing systems seem to have been one of the last elements to develop.

But there were things that these civilizations did not share. They did not all have a highly developed technology. The Maya civilization in Mesoamerica for example, even at its peak had no wheel, no draft animals, no metal tools, and relatively simple forms of agriculture in most cases. Nevertheless, it had the highly developed sociopolitical organization of the other early civilizations. It had excellent astronomers, and the Maya calendar was more accurate than that of the Spaniards who conquered the Maya in the sixteenth century A.D.

Let me offer now one of the current hypotheses for how man in the Near East began agriculture, where, perhaps for the first time, man had started farming anywhere in the world. All hunting and gathering peoples, as Doctor Lee has already said, tend to maintain their populations below the level at which they would exhaust their wild resources. One of the mechanisms for this is emigration to new areas as populations rise. Archaeological data show us that populations grew and emigrations took place throughout the Pleistocene, but by 10,000 B.C. all environments of the world, except for the extreme Arctic and Antarctic, had been occupied. That is, although populations were still increasing, there were no vacant lands left into which they could emigrate. At the same time, sea levels all over the world rose as the glaciers of the late Pleistocene melted, and this converted many lower or coastal river systems into tidewater estuaries. These coastal estuary systems were extremely rich in fish, shellfish, and marine food resources—invertebrate foods.

Around the end of the Pleistocene we begin to see the first permanent coastal settlements of people utilizing marine resources much more heavily than man ever had before. In inland areas we see a great increase in the variety of foods which man was eating and which can be found in the archaeological sites. In other words, as the world began to fill up, man, who had formerly been primarily a hunter, turned increasingly to plant, invertebrate, and marine foods. Many of these foods were ones which could be collected by women and children and many of them were abundant enough in sufficient localities so that people could now live permanently in one place all year.

This did not solve the problems of population pressure as population continued to increase, especially in those areas where the early sedentary communities are found. During this period, all over the world, we see a tremendous variety of experiments with ways of increasing the local food supply. This included the working out of technologies for making edible some types of plants which had previously been inedible, and turning to types of food which had previously been ignored.

One of the experiments made in the Near East was to settle permanently in areas where wild wheat and wild barley grew naturally. In some parts of Lebanon, Syria, and Jordan, wild wheat and wild barley, under natural conditions will yield 500 kilograms per hectare. This means that in three weeks of harvesting, a family of four could harvest a year's food supply of wheat and barley, and by 8500 B.C. there are permanent villages in that part of the world with very well-developed storage facilities; in the archaeological remains of some of these sites are the carbonized seeds of wild wheat and barley.

One current theory is that agriculture began not long after this in the drier areas around the margins of these favorable habitats, in areas where one could only raise the productivity to 500 kilograms per hectare by deliberately planting grains. By 7000 B.C., we begin to find the first carbonized cereal grains which show genetic changes from the wild form. This is accompanied by a developing technology of storage facilities, of flint knives for harvesting grain, ovens for roasting the grain, and stone grinding tools for threshing.

During this period, in a relatively short period of time, the village became the dominant form of settlement over most of the Near East. The early villages, so far as we can tell from archaeological data, were composed of houses occupied by nuclear families in which the family was the basic unit of production and storage and in which ancestors were buried under the floors of the houses; in many cases their skulls were saved. The suggestion is that, in contrast to most of the hunters of the preceding period, we now had a form of social organization which crossed many generational lines, in which dead ancestors were thought to continue to take part in the activities of the community.

The burial grounds indicate that families had become larger and that probably in this new agricultural society there was selection for larger numbers of children than previously, because there are so many tasks in an agricultural society that can be performed by children (who are really a burden in hunting societies).

At this stage, every man was his own craftsman; that is, every household was capable of the kind of craft activities it needed to be self-sufficient. But by 4500 B.C. we begin to get the first evidence of villages which are specialized in craft activities.

I will just give you one example from the neighboring country of Turkey. By this period, near Lake Van in Turkey, there is at least one prehistoric village located on a source of obsidian or volcanic glass which it was processing to use as knives or blades for export.

During the period between 7000 B.C. and 3000 B.C., populations in parts of the Near East increased 60 times. Villages grew larger and public buildings appeared in them. In the burials, one can distinguish a hereditary elite which is buried with exotic raw materials imported from very distant areas.

As population increased, agriculture was intensified and came to include systems of irrigation; some villages in good agricultural localities had defensive walls or ditches.

By 3500 B.C., competition for scarce strategic resources had become so great that one sees a complete difference in settlement. In some areas the number of actual villages is greatly reduced, and the population seems to have been drawn together into a city. At this time, the craftsmen were pulled out of local villages and placed in wards or sections of the city. Sometimes the cities themselves are spaced so regularly, at standard distances from each other, that they seem to form the hexagonal patterns which geographers call central place lattices; that is, the city had become a service center for a very large area of rural settlements and had taken on all the administrative and craft functions which formerly had been found in the villages. In the course of development of the temples in the center of some cities, one can first see the appearance of small residences which are evidently those of the caretakers of the temple, and according to archaeologist Robert Adams it is these small residences which, in the space of 500 years, become the palaces of the rulers of the city.

At this point, the first writing begins; and the earliest Mesopotamian writing seems to consist of accounts of incoming and outgoing agricultural products and animals. In other words, by this time such enormous quantities of information had to be processed for society that there was a professional class of scribes, and a way of information storage on clay tablets had been developed.

If we can consider these ancient societies as populations exchanging matter, energy, and information with their environment, it seems clear that later institutions that developed, which characterized civilizations, were almost always ones for information processing of different kinds. These institutions were initially established to serve the system by processing data faster, but as so often happens, once the institutions were established, they became self-serving, so that they continued to expand and to place demands on the system for their own support.

At this time, we have reached the bureaucratic civilizations of 2500 B.C., many of whose characteristics are not very different from our present-day civilizations: they not only had a system of laws and precedents and the recording of legal activities on clay tablets, but we can even read on the tablets of people complaining that the lawyers took their money and did nothing.

This is the basic sequence as revealed by archaeological data. If I had to pick three questions that ought to be answered in the future, these would probably be the most important:

First, why did certain people in the world begin producing their own food by agriculture while other people found alternative ways to meet the problems of population density? Did it have something to do with the genetic plasticity of the plants themselves in that region? How was it related to social organization and division of labor? Was it an effort to minimize the difference between wet and dry years, or was it the result of a pattern of intensive harvesting of wild plants?

Second, how did stratified societies arise? Was it through investment of power in a leader for protection; or investment of power in his role as a religious specialist; or for more efficient data processing and decision making; or was it related to more intense competition for increasingly scarce resources?

Third, how did state-supported technology arise? This is an important point for our purposes, because it was the forerunner of applied research for technological answers. Was it to attack problems of increasing the food supply, or for developing weaponry for protection against competing groups? Was it to produce luxury items for the nobility, or for trade, or for the gods, or for all these reasons and more?

CRICK: I think it would help if we reformulated the problem in front of us. The problem we have is to estimate the likelihood of intelligent life on other worlds, and in particular the distance of the nearest civilization.

There is a subproblem which we should also consider and that is, if life had started on Earth a second time, how long would it take to evolve beings with the intelligence of ourselves?

The formulation which was put on the board by Professor Sagan (equation 1) was a steady-state one. The second problem I am considering, the subproblem, will be best tackled by expressing it as a rate process. For all the different steps which have to happen in sequence, which were shown by equation (1), we estimate the most probable time for each step. In this formulation, we then add them together as probable times.

From this point of view, it seems to me (and this is the point I want to make) that there is a very large difference between the presentation we heard from Doctor Flannery and that from Doctor Lee. Because if any one step takes a relatively short time and we have evidence that it happened independently in various places, then our estimate of the mean time that accrued will be a short one. Even if our estimates of the time for the subject covered by Doctor Flannery's talk was wrong by a factor of 100—in other words, if it took a hundred times longer than he says—this would not disturb us very much.

When we come to the subject of Doctor Lee's talk, on the other hand, we are considering the time from the simplest organisms up to man and this is a very long time. Therefore, if we make an error in the estimate of that time, we could easily get up to figures which are longer than the age of the universe.

Therefore, although I count both papers fascinating, I conclude, that the discussion should be concentrated more on Doctor Lee's talk than on Doctor Flannery's, and I would like to make a few preliminary remarks about Doctor Lee's talk.

I propose to omit a certain technical problem. This is the problem of how, in evolution, we went from microorganisms to what are called eukaryotes—to the simplest higher organisms, of which yeast is an example. I propose to omit that, and I propose to consider the period from yeast to man.

We can, I think, assume the following: that once you have single cellular organisms with the molecular apparatus that you find in yeast, there is bound to be an evolution to organisms with many cells and to many different organisms of increasing degrees of complexity as time goes on. However, this does not give us an infinite or extremely large population of different organisms in the following sense: if you consider the insects, for example, they are in some sense caught in an evolutionary cul de sac, although they are extremely successful animals—because they cannot evolve to be very big owing to physiological requirements about their supply of oxygen, and indeed no insect has gotten much bigger than a cockroach. Therefore, we shall have to consider large classes of animals and not just a certain number of species.

The strong impression that I got from Doctor Lee's talk was that in order to evolve intelligence, there is a fairly large number of different kinds of criteria that had to be satisfied. I start from the assumption, in which I understand I am supported by Professor Minsky, that in order to be intelligent and to form concepts, you must have a reasonably large number of elements making up the nervous system, in this case a large number of nerve cells. There appears to be a limit to the degree of miniaturization which you can effect—you cannot have too small a nerve cell, apparently. Consequently, there is a minimum mass of brain that you must have in order to be intelligent.

This would explain quite easily why the insects have not been able to solve mathematical problems, although they have a social life and some of them look after their young, and so forth. But for mathematics I think they are too small, so the insect need not excite us.

When we come to the reptiles, on the other hand, the reptiles are very large and very various. When they appeared they were very successful. They did not have particularly large brains but we don't know any reason why they might not have had bigger ones.

On the other hand, most reptiles, for example, do not look after their young because of the way they develop, and for this and possibly other reasons, one can see that the reptiles, too, are not able to do mathematics.

Birds, which are very intelligent creatures in a zoological sense, necessarily, because they have to fly, have to have small brains.

There are a few other examples. In the case of the giant squid, which is certainly large, I think at least the smaller ones go in shoals and, therefore, communicate with each other. They have their arms with which they can manipulate things. We have to ask what it is they couldn't do, and I would also direct your attention to wolves which hunt and go in packs. They don't have hands, and so on.

I don't want to elaborate the argument. The argument is that many, many different properties would be needed in order to form an intelligent animal.

The second point I want to make in this context is, Did it happen only once? We are, therefore, in the same logical situation as we were earlier when I expressed my opinion about playing cards, that you cannot estimate the mean time of something happening from a single event unless you have some prior knowledge.

Therefore, I come to the conclusion that since this process, anyway, appeared to take a long time, it is very difficult for us to obtain any estimate at all of the time it would take if, on Earth, we had to go through the whole process again, and intelligence had to arise a second time.

Therefore (and I now repeat myself), what Doctor Flannery said does not give us a problem; but what Doctor Lee said does give us a problem because we cannot use an estimate of time from one occurrence, and we do not yet have a deep enough understanding of the process to make, even for this, a reasonable estimate of the mean time for the origin of intelligence.

LEE: The relative proportions of time from the origin of life to the origin of intelligence on the one hand and from the origin of intelligence to the origin of civilization on the other, are each about 99.9 percent—from 4 billion years to 4 million years, and from 4 million years to, let's say, 10 thousand years.

So, in the perspective of the origin of life, the origin of intelligence is no later than civilization is in the perspective of the origin of intelligence.

CRICK: I think it is the absolute times, not the ratio of times, that is my point.

LEE: On this whole question of could it happen twice, given the tremendous species diversity and the tens of thousands of ecological niches that have been occupied by life forms on this planet, I get a strong subjective impression of the inevitability of the adaptation for intelligence. It will eventually be achieved by some life form, given just the experimental, exploratory nature of adaptations on this planet.

GOLD: I would like to refer to another possible driving force for the development of intelligence that has not been mentioned, although it was peripherally indicated, and that is this most interesting possibility that there existed another humanoid form which did not intermix, which could not intermix with our ancestors but which coexisted for perhaps as long as 2 million years. If that story is correct, and there seems to be some strong evidence in favor of it, then you must think of humanity having lived for this very long period in fierce competition with another race which almost certainly took interest in the same areas of land and the same circumstances, and probably hunted the same animals, and so on, and would, therefore, be permanently a severe competitor.

If that is true, it seems to me that the race which perhaps was physically inferior, which is what our ancestors seem to have been, would have been crippled in an attempt to outwit the competing race all the time, and if that existed for nearly 2 million years, that would have been a very important selection feature in a drive toward superior intelligence. Perhaps such fierce competition is, indeed, a necessary thing to create our level of intelligence.

MORRISON: Especially when Doctor Crick rests his point on

somewhat agnostic grounds, one has to agree that one event can't be a statistic. It is very hard to beat that.

It is also certainly true that the repetitive summation of the time scales means that the longest gap which we see, which is the 4 billion years perhaps between the first nonorganelle cell and ourselves, is the most serious in terms of error. But I think there are two points about the second coming to which I should like to refer. One is already to some extent implied by Gold's proposal, at least as a very general proposition—it is very hard to fill a niche twice in the presence of existing species, so we need not be surprised if in the phase space of species around intelligence there is a big gap, because the presence of one will repel all others around it, for obvious reasons.

That is true at every level of the evolutionary process, so I would like to ask what is the incidence of forms not close in their plasticity to the origin of intelligence, as were the primates of Doctor Lee of 5 million years ago? What was the incidence of other mammalian forms (there may be only one of a few classes, but that is still a pretty fair statistic) which looked like our ancestors, not in the beginning of the Pleistocene but in the Eocene, 50 million or 100 million years ago when they were, I suppose (and I am no primatologist), little lemurian fellows that hung onto branches and weighed one pound and yet had the potential of intelligence.

I look around the mammalian families and I think there are several that look favorable to me, if we were not here and they were let go for a time as long as the lemur to man. I mention as one example the omnivorous, hand-using, cute and responsive raccoon.

MINSKY: I would like to go along with Morrison. There are several candidates—the whale, the dolphin, the cephalopods. I don't know how old they are in their present form but they have pretty well developed protobrains, at least. As Doctor

Crick mentioned, some of the birds are pretty smart. But if the ecological situation changed so that they could come down and didn't have to fly, like the ostrich, well, the ground is pretty cluttered and birds have spent their histories flying.

Because of this, I would like to consider the concept that there is an ecological niche for intelligence. I don't know if anybody really proposed that, but it seems to me that no matter where you are, if you could afford the extra weight and metabolism, intelligence would be good. But I liked very much Lee's point about the danger of having a little intelligence. Intelligence is experimental. In order to survive in an environment when you are not very smart, you must rely on a highly, carefully de-bugged instinct. You must not try experiments. Just as in the case of mutations, intellectual experiments are almost always fatal, and so we have Lee's point that in a social environment the children can try experiments and, much more important, I think, they can be told.

That is why one of Doctor Lee's later remarks puzzled me. He seemed to be disturbed at the question of why the non-technological civilizations, once having developed language, went on to tell stories and develop a highly intellectual, non-technological culture. There is, it seems to me, a technical point. Once one develops language in the theory of confrontation, one achieves what is called universal computational ability. And it is very difficult, in fact it is impossible, to make a computer that can do any nontrivial calculations and that is also not capable of doing all nontrivial calculations. It is a curious tech-nical point, just given enough memory. So, it seems to me once you have developed language, one of the prices is that you have to develop humor and all of the other social games for passing the time, and it takes a highly developed civilization then, and a very strict set of mores, to prevent doing the nontrivial prob-lem solving.

MUKHIN: Doctor Lee defined intelligence as a means by which an organism deals with complicated situations, and social life was also brought into this. On the other hand, Professor Crick has said that intelligence is characterized by the weight of the computing system. It seems to me that we must introduce certain correcting factors into both definitions, qualify them. The fact is, we know that very many organisms can deal with complicated situations and have a mode of existence similar to a social life. It is hard to believe that anyone here would claim that insects possess intelligence.

At the same time, we know brain systems that are bigger than ours—take whales—and it is hard for us to estimate the degree of their intelligence, although undoubtedly these animals are intelligent. Therefore, it seems to me that in examining all these possibilities, it is extremely important to abide by strict definitions, although they are not our prime concern. Therefore, my question to Doctor Lee is for a more exact definition of intelligence and complicated social situations.

LEE: I will try to answer that briefly. A typical example of a complicated situation that a primate might face is one involving group-living animals arranged into dominance hierarchies; animal 3 has got animals 1 and 2 above him and animals 4 and 5 below him. The female is in estrus, he would like to mate with her, and has to work out the best strategy of appealing to allies between 1, 2, 4, and 5 in order to strategize his way to the desired goal, namely, to copulate with the female.

These dominant hierarchies are complex, nonlinear, involve coalitions and involve many species of primates in very complex minute-to-minute or second-to-second changes in behavior, which I see as a very interesting readout for the development of complex behavior.

E. MARKARIAN: I would like to say a few words about Doctor

Lee's remarks. I listened with great interest and I agree with much of what he said. I agree specifically, that the concept of adaptation is a key concept in understanding civilization and its genesis.

In this connection, however, I would like to make two critical comments.

First, at this symposium the problems of genesis of human society should be discussed from quite a definite point of view which could direct attention to the construction of an abstract genetic model of civilization. This genetic model of civilization must embrace its possible extraterrestrial manifestations as well.

Second, in order to make the concept "adaptation" fulfill its important methodological function in the given theoretical context, it is necessary first of all to determine the specific features of human society as an adaptive system.

Theoretical syntheses of fundamental concepts and ideas often arise through a new, extended interpretation of earlier concepts developed in narrower fields of learning. The concept of information and control, for example, arose in the social sciences and only then, following the emergence of information theory and cybernetics, transcended those boundaries and acquired the new meanings that they have today in these fields.

The reverse process also seems possible when concepts born in the natural sciences may acquire a much broader meaning on the fertile ground of the social sciences—for example, the concept of adaptation.

This is a case in point. However, referring human society to adaptive systems would make it possible to examine human society in a broader perspective with other self-regulating systems, and probably the most complicated stage here must be the problem of establishing the specific features of human society as an adaptive system. This leads us to the question of how, by what sociological generalization, can we formulate the adaptive behavior of human society?

I think that human society can be referred to a special class of systems which we will call adaptive-adapting. This definition has the purpose of expressing the specifically active and transforming character of human activity in its relation to nature.

The specific elements in the behavior of human society consists of a general adaptation to the environment through a counter-adaptation of the environment to human society to meet its needs through a purposeful influence. In other words, adaptive-adapting system is a term that reflects two aspects of the behavior of human society which is both adaptive and adapting. This aspect of human society may possibly be extrapolated to extraterrestrial societies.

It is true that the element of "adapting" activity is also typical of some animals. However, the adapting activity of the animals, first, is genetically programmed and strictly specialized; second, with rare exceptions, it involves no technology.

It is the concept "culture" ("civilization") which in modern science expresses integratively the system of means and mechanisms which allow the possibility of working out a universal, nongenetically programmed type of "adaptive-adapting" activity.

From this point of view the cosmic essence of culture should be regarded as an ability of living beings united into some stable collectives to create a system of means and mechanisms potentially not preconditioned by the biological type of organization. It is due to these extrabiologically derived means and mechanisms that the activity of the members of the sociocultural system is stimulated, programmed, coordinated, fulfilled, and provided for.

In this theoretical prospect civilization must be regarded as a specific way of "negentropic ordering" of matter. It is quite

possible that this understanding of the phenomenon of civilization may become a theoretical basis for its functional definition.*

The transforming, adaptive-adapting character of a system's
activity on entering the path of civilization poses a very important practical problem, a problem involving the relation between
entropy and negentropy in the interaction of these elements.

As you know, negentropy effects developed by civilizations
and other self-regulating systems can only be local in their
character, and a decline in the entropy of a social system takes
place through a rise in these tendencies in the environment.
What, then, I ask, are the limits imposed on this process? This
is the problem that is bound to arise for any highly developed
civilization because a natural result in the rapid progress of a
civilization is the rise of entropic processes in the environment.
Therefore, the problem of the adaptive potential of humanity
is a social problem of outstanding importance.

In conclusion, I would like to ask Doctor Lee the following:
I would like to know his opinion on what are the main channels
and means whereby animals accumulate acquired experience in
their lifetime and transform it into collective experience.

LEE: The reason I didn't use the word "culture" in my presentation is that culture, in the sense of a nongenetically transmitted repertoire of behavior that is capable of accumulation
through tradition, is observed in the nonhuman primates as well
as in the human primates. There is some capacity for learned
accumulation of behavior in the monkeys and apes and no evidence, I believe, of the inheritance of acquired characteristics
that we were speaking about earlier. This does not represent a
qualitative difference between man and animal but only a quantitative one.

* These problems are discussed in detail by the author in the following
works: *The Problems of Systematic Study of Society,* Moscow: Zhnany,
1972; *On the Genesis of Human Activity and Culture,* Erevan: Publishing
House of the Armenian Academy of Science, 1973.

DISCUSSION

ORGEL: I would like to go back to an earlier subject and produce an argument which I hope will show the lack of wisdom of making predictions in the absence of any background information.

Let us consider that some event indeed has occurred on the earth. Let us ask what is the probability that that event, B, will occur on a planet similar to the earth. In the absence of background information, the nature of event B is really quite irrelevant, because that merely provides background information.

I should now like to take three events: first, an atmosphere has developed on this planet; second, life has developed on this planet; and third, a discussion between Doctor Crick and Doctor Sagan has occurred on this planet. In the absence of background information, we are obliged to admit that these three events can be predicted to have equal probability on the planet. Any theory which makes predictions of this type I feel is not a theory on which we would want to base any actions.

SAGAN: I quite agree. But we *have* background information. If the planet has a sufficiently large mass and not too high an exosphere temperature it will likely have an atmosphere. But the *a priori* probability that a conversation between Doctor Crick and myself should occur—much less, that we should disagree!—is minuscule.

MC NEILL: It seems to me that with your master formula, equation (1) with f_l, f_i, and f_c, this choice of critical stages, as it were, is perhaps a little arbitrary. I was thinking of how you arrived at your number and the possible frequency of cosmic civilizations by a process of multiplication of negatives. The number of stages that you choose to set between the creation of the earth, the condensation of the solar system and the achievement of contemporary civilization is really arbitrary. You have three *f*'s but you could just as well have five or six of them. For example, you could distinguish the division between plant and animal, and then you get another filter with another degree of probability.

SAGAN: Or the evolution of eukaryotes from prokaryotes.

CRICK: I think the evolution is subdivided into stages and to identify the stages for which the time is small, we have a reasonable estimate of what it is; but for the stages where the time is long we do not have an estimate because those are the difficult ones. So, for example, it could be argued that once the mammals have started to evolve, it would perhaps be the sort of order of time that we see, plus or minus not too big a factor, before we got to something like man, in which case, if that were a tenable point of view, we needn't worry too much about that stage.

I would say that one of the points of this meeting is to identify which stages are the ones which we are really ignorant about and which we can leave to one side. I would agree with your analysis—

MC NEILL: But that is skewed by the number of f's you choose to put in. Isn't that correct?

CRICK: Yes.

SAGAN: Not if each f has a value of one. Then you can put in as many f's as you want.

CRICK: If his formulation is right, if you divide an f into two parts, then it is simply the product of those two f's.

I want to say also that you want to establish an uncertainty about the value of the f and you want to divide the f's crudely into ones that you are reasonably sure about (those you don't have to worry about) and the ones which there really is uncertainty about, of which I would claim the origin of life is one.

MINSKY: My question isn't to the point of answering this particular question on which is the weakest link. My question is what our recommendation is supposed to be in two possible cases. First, if N [equation (1)] is low, then I think if we are going to listen we have to listen to a lot of stars; but if we can convince ourselves that N is very high, then we can erect 20 antennas and listen to the 20 nearest stars and be pretty sure that one of them would succeed. So the question is at one end how many antennas to erect and at the other, if it turned out that there was just one very weak link, like the development of mitosis or whatever, then we could also recommend doing more research on that particular facet of biochemistry. However, we always concentrate on what we think is very important and poorly understood.

CRICK: On the question of search, which we probably ought to defer until later, then we have to adopt, I think, the point of view of a well-known story which I first heard told by Tommy Gold in the Cavendish, it must be twenty years ago—the story of a man looking for a key that he had lost. You lose something in the street and there are a few lights. You don't look where it is dark, you look where it is light, because it is all you can do at the moment.

The analogy is very precise in this case, the light being the reasonable range at which we can search. And it seems to me we are bound to come to that conclusion—that you look in the light area because you can't define with any probability the amount of light and dark.

SAGAN: But you can have different strategies for different contingencies. You can take one strategy, to look only for nearby dumb civilizations, or you can take the strategy of looking for faraway, smart civilizations—Kardashev's Type II and III civilizations—and those require a very different search strategy.

CRICK: I wouldn't want to say there was only one strategy. I would agree with you completely, and I was going to say that a proper analysis (which I haven't thought through) comes to the likely cost and return. This is what we need the f values for. But in fact this is the difficult part of the problem; it doesn't solve the problem but perhaps it puts it in a graphic way.

SAGAN: After all, there is going to be a very large number of probabilities if we could identify them all. You can imagine an equation which has a hundred f's.

MCNEILL: At least.

SAGAN: Or a thousand f's. And now you could take an optimistic point of view and say: I will just attach a 90 percent probability to each of those factors. Well, when you get done multiplying 0.9 by itself a thousand times you are going to get a very small number.

MCNEILL: This is why the omission of these f's bothers me. You can't really say an f has a unit probability—I don't think you can.

SAGAN: Sure you can.

MCNEILL: You have to get a probability of less than one for all of these factors.

SAGAN: I don't agree. There is no question that there are a great many individually unlikely steps which led to the development of our technical civilization. But are there not many other sequences of steps that would lead to a more or less equivalent civilization? Is there a kind of quasi-ergodic theorem in the evolution of civilizations so that you will always wind up, by some path or other, in that particular area of intellectual development? If so, it might be perfectly acceptable to have many of the f's equal unity. It is on this point that some of us disagree with George Gaylord Simpson, who has kindly summarized briefly his views for us in Appendix B.

SHKLOVSKY: Could Doctor McNeill repeat his problem?

MC NEILL: In equation (1) there are three f's. It seems to me that this choice of three is very arbitrary. By adding additional f's—that is, a step between vegetable life and animal life as a separate f, or the step between the development of intelligent human beings from hunters to food producers, and then from food producers to the civilizations we heard about today, and then a civilization that had, let us say, higher religion as a stabilizing factor in its social structure, and then a civilization that developed technology as the Western World did in the seventeenth century, etc., etc., you could get any number of interstitial steps, and each one with a probability of less than 1.

This is my question to you. If you put in a number of f's, then your probability begins to decrease and decrease and decrease. The counterargument, as I understand it, is that there may have been alternative paths to a comparable result.

GOLD: If there were not, then we would say the search for extraterrestrial intelligence is totally hopeless.

SAGAN: If there were not many paths to a functionally equivalent result, N in equation (1) would equal unity.

ORGEL: As an analogy, consider the possibility of getting from Cambridge to Erevan. We might decide that that was very difficult, that the probability was 10^{-4}. However, you might say: How can you take such a big step? Wouldn't it be better to split it into three steps, from Cambridge to London, from London to Moscow, and from Moscow to Erevan. Then we would say there will now be three factors, so the probability would be less. This argument does not hold because when you look at these three factors separately, you find that there is a somewhat larger chance of getting from Cambridge to London than from Cambridge to Moscow, and when you multiply these three factors together, you come back to exactly the same result.

However, suppose now there was also a route to Erevan which goes from Cambridge to London to Budapest to Erevan; then the net probability of arriving in Erevan is greater, and it is necessary to add the probabilities of alternative pathways. It would modify the results but still it wouldn't change the numbers very much if you considered only two pathways instead of one.

What I think people like Carl are trying to do is to make a guess of the total Cambridge to Erevan probability and they claim (and I don't know that they can do it) that they can justify it by breaking it up into a few separate pieces. Is that fair, Carl?

SAGAN: Perfectly fair. Also the alternative pathways are important. If someone managed with great difficulty to accomplish the Budapest-Erevan flight, he might conclude it is almost impossible to reach Erevan. But the pathway via Moscow makes arrival in Erevan much more likely.

ORGEL: So, that is why it doesn't make any difference how many f's we have.

MINSKY: If you make many steps, then the product ends up the same way—the probability of exactly making man is very small. But there is not just one pathway to the equivalent of man.

PLATT: I want to claim that there are some steps whose probability is very close to unity. I call some of these lock-ins, because one has a cycle of events which are self-maintaining. There are some steps in evolution where, when you get over a certain threshold, you are almost certain to fall down into the basin and it doesn't make any difference particularly by what path you go—you fall down into the basin anyway.

Probably a case of this sort, in my opinion, is speech. Once you have passed the threshold of speech you generate language which, as Minsky says, is a universal computer and can eventually do anything, including creating humor and myth and stories, and so on. From that point on, the path is relatively certain to a high technology.

CRICK: That was the point I was trying to make earlier. I think this is why it is useful to think in the other formulation in terms of time. Let us take the concrete case of f_c. The impression I had from listening to Doctor Flannery was that once you got to this particular stage, then the time to form a civilization was a relatively short time.

Consequently, if you make an error in estimating a short-time-scale step—which is essentially what Doctor Platt was saying: he said you will get there in the end, he didn't say how long it would take—it doesn't matter if you are wrong by a factor of 10, because we have factors of 10. Those are the steps we ought to agree to leave on one side. It would seem to me equally if you get to the stage where you have got people doing mathematics, which is one of the criteria I use for intelligence, incidentally (it is an arbitrary one but useful), you are bound to get to a high technology; and since the time is very short it doesn't matter if it was a bit longer somewhere else.

So, what matters are the steps for which an error in estimation makes a big difference to the actual total time. I therefore feel one of the objects of this meeting is to identify which of these substeps—and I agree absolutely with what was said earlier, that there are many more—we can ignore because in that terminology the probability is going to be 1. If we want to approach the problem in this direction rather than from the point of view of the man just looking where the light is—then you have to go at those steps which took a longish time. About these steps there is real doubt (because they were unique events) as to whether our time is typical. My view is that the origin of life is one of those, but I don't know enough about the whole course of later evolution to divide it up into substeps and say which ones, as Simpson would say, are OK and which ones are not.

Once you have got to the mammal, then it doesn't matter if you take twice as long. It is irrelevant to the calculation, so why do we fuss about it?

SAGAN: Let me say what Simpson's point of view is. He has a paper called, "The Nonprevalence of Humanoids" [see Appendix B]. The principal argument of this paper is that there are a large number of individually unlikely steps which are required for the evolution of man, and that the chance of the random recurrence of this sequence of steps is so small as to make, I would say, the existence of mankind impossible.

PLATT: He is talking about man and not intelligence.

SAGAN: And that is precisely what I think the criticism of Simpson's view is—that there may be many, many other pathways to an organism which is functionally equivalent to a human being but which looks nothing like a human being. Simpson, by the way, believes that the really difficult steps in the origin of man occurred late—Mesozoic and more recent times.

GOLD: But in that case, being the first to appear on the scene with that level of intellect, you would have to suppose that the evolution to a mammal was the limiting factor, because, after all, you had a very long period before when all kinds of other animals existed and did not so develop; so the probabilities for them must have been small. For all the very many reptiles that existed for millions of years, the probability for their developing into intelligent beings must have been very small. There has to be something very specific about mammals. It is not clear to me why that should be so.

MORRISON: In the first place, there are only 300 million years of land life.

GOLD: Animal land life.

MORRISON: That isn't a very large time. Plant life much before that—450 million years, perhaps—but for animal land life, 300 million years. That is only one-fifteenth the age of the earth, so I don't even say that this was a necessarily very difficult step. It is really only a short time. Another 300 million years would only be a 7 percent effect and would double the possibilities, which are interesting.

GOLD: But sea life was developing in almost the same span.

MORRISON: No, not at all. Dolphins and whales are not sea animals. They are returned land animals and showed interesting plasticity. But I would like to raise the following question which has always puzzled me—the shape of fish. If you ask how difficult it is to evolve a creature that feeds upon fast swimming fish in the deep ocean and manages a top speed of 30 miles an hour, which means you have to be quite good size—6 feet long or something—I am sure that by simple analysis it is not very easy to evolve such a beast; but three such species have evolved out of completely different sets of steps.

The biologists, of course, don't regard them as the same but to any observer looking at them and looking at their performance, they are about as similar as a very good automobile and a little old flivver which isn't the same car but is recognized as doing the same thing. One is a reptile, the Ichthyosaur; one is a large fish 100 million years later; and one is a mammal, the dolphin. They all have about the same top speed. The endurance and the electronic installations of the dolphin are much better, but they all do about the same thing. They all lived on small ocean fish and they lived very successfully. They came by three different routes and their function is the same.

PLATT: You would say that anytime you get extensive convergence of evolution along different lines like this, you must be dealing with an almost certain process. So what we have to look at are those steps for which we have no evidence of convergence. Where we ought to look hardest is where there is a possibility of convergence.

SAGAN: It is really very difficult because in some cases convergent evolution wipes out your predecessors, so you may not be able to find the convergence. For example, convergence in the origin of life: Even if the chemistry of the earth were proper for the origin of life today, you could not have it recur today for the reason that Darwin recognized, namely, that the little beasties would be eaten up by the ones that are around today.

On Doctor Morrison's point that 300 million years is only 7 percent of the age of the earth, I again point out that the origin of life likely took less, making the origin of life a probable event.

I would like to ask Doctor Hubel about convergent evolution in the eye.

CRICK: The squid and the mammalian eye.

HUBEL: They are very different.

MORRISON: No. They work the same way, don't they? They have the same performance.

GOLD: Allow me to say a few words on this; I have looked into it a little bit. If you think that the two eyes were developed by totally different routes (I mean that the eye of the octopus and the mammalian eye or reptilian eye do not trace back to any common ancestors until you come back long before the evolution of any eye, so they were all developed quite independently) and if you look at the detailed design of the two eyes, you must conclude that they are remarkably alike. In fact, they are so much alike that I would venture to say, while I do not know what some creature on some other planet might look like, that if it is fairly highly evolved and if it is based on the same basic biological mechanisms as we are (it no doubt will have totally different body shape, and so on), there is a high probability that its eyes will look rather similar.

The octopus eye has the mechanics of an iris much the same, it has the retina the other way up so that you can tell the difference, but even under the microscope it still has an awful lot of structures that look similar in detail.

MORRISON: And it has an embryological history just quite the opposite.

HUBEL: Some things are the same and some are very different. The receptor cells hyperpolarize when you shine light on them in the vertebrate eye; they depolarize in the invertebrate.

GOLD: I would like to attack our previous chauvinism in discussing the development of intelligence too closely related to the development of human intelligence. For example, I could sort of visualize a discussion like our present one taking place on another planet, with different circumstances, and I could quite easily see a circumstance where the little beasties discuss how, in order to develop intelligence, it is necessary to be extremely small, to have no individuality, no individual personalities, perhaps social organization in this particular example.

Take the case where you have small insectlike animals, and you know insects that live in large groups like ants do communicate with each other—the ants do it by wiggling each other's bellies. That's a very slow matter, I agree. But suppose that some other insects have developed communication by making electrical contacts with each other.

Suppose that they have developed electrical contacts to communicate with each other and they got into some very outlandish possibility that there is, say, direct nerve contact to be made externally. Suppose that they initially are required to do some major construction work, as ants do, civil engineering. I could foresee that intelligence would develop as a result of their linking up in particular patterns with each other, first by a few and then by the millions, and that they would gradually evolve patterns in which they could connect themselves to each other so that the group as a whole could do very elaborate computations. And then they would argue, if they were arguing this case—

SAGAN: He, not they.

GOLD: He would then argue that what it takes to make superior intelligence is first, small nerves so that very many can be connected to each other in complex patterns; second, that they must be alike or at least in subgroups they must be alike so that no errors are introduced, just as you want transistors to be alike.

Then he would argue that no single organism would be as satisfactory because when it dies—and they would understand that eventually any particular chemical arrangement is going to go wrong—all its memory and learned capability is lost. But instead, in this tribe of ants, when anyone drops out dead, then another comes along and pushes him out of the way and connects himself at that place, and it merely means a temporary interruption of one chain and there are many redundant chains so nothing is lost. The organization therefore lives forever and they would take that as an enormous advantage for developing superior intelligence.

What we can conclude from this is that we must think very widely as to what it takes to develop intelligence and not take us so much as a model of what is necessary.

CRICK: One of the weakest links, in my opinion, is the small size of the neuron on which we have no general knowledge at all;

in fact, we can discuss that if you miniaturize things, a neuron on a molecular scale is very big; I might ask Doctor Minsky what is the minimum total number of elements you need for intelligence?

MINSKY: I think, technologically, you can probably get 10^{12} memory elements into a given millimeter within the next 50 or 100 years, so that one could, in principle, have a brain that small. But I don't know if you could construct it genetically.

I wanted to make two remarks which are in the nature of chauvinism. I read a paper by Isaac Asimov in which he constructed an argument in favor of the likelihood of the human form which was pretty good. For example, one argument was that the important, powerful sensory centers ought to be connected in the brain by fairly short paths so that the animal could have short response times. He had an argument for bilateral symmetry and had the usual argument that a good pair of hands in front of the body, where the eyes can see what it is doing, is very good. The octopus has a bit of a problem; in fact, it doesn't coordinate its hands and eyes very well.

So, Asimov gives fairly high probability for a large number of humanoid characteristics.*

On the question of exterrestrial intelligence, I think that, at least from what we know about intelligence now, there are certain aspects of intelligence that would be necessary even if one disagreed about details. The ability to use knowledge is important in intelligence. It is probably very important to be able to get knowledge and transmit it. It is not very important to be able to generate it. The ability to create new ideas is very unimportant if you are in a culture. None of us generates very many ideas, or at least we don't recognize our debt—that with each word in a language we learn comes a concept from our

---

* Editor's note: See also R. Bier, Humanoids on other planets, *American Scientist* 52 (1964): 452.

culture. It is probably easier to communicate with a Jovian scientist than with an American human teenager.

It is the nature of intelligence to be able to build up an abstract knowledge, and it is very much like another point I made about the universal computer. Once you can store knowledge and create new procedures in your head, then you run and use those to modify and utilize other kinds of knowledge. What you think will depend much more on your culture than on how your brain works. In other words, I think there is an enormously important type of convergence that has to occur. The ability to run various types of computer programs in accordance with instructions that you receive from your culture is the key element of intelligence, and the alienness of the aliens will be the peculiarities of their culture.

But I think there will be a convergence just as in the necessary objects of the eye that forces this convergence. You have to have something like an iris, you have to have a lens, you have to have a retina, and the thing shapes up so that there aren't very many ways to design an image-forming eye. I think the same thing is going to be true of a concept-manipulated brain. It has to be able to deal with arbitrary strings of symbols. It will have to be able to form certain kinds of associations. It will have to be able to match two strings of symbols and see the different lists. It will have to have temporary storage that can work backwards, that can bring items down in inverse order to the way they went in, pushed down with the program.

So I think that things are fairly hopeful. The science fiction idea of aliens that are so alien that you can't talk to them, even if they are intelligent, I think is unlikely if they are scientists. One can always imagine science completely alien, but one can't imagine scientists not knowing what I am talking about.

PLATT: How many neurons does it take to create intelligence? Does anybody have any idea of the minimum number of neurons—is it 10^{10}, is it 10^6? Is it possible that we use ours very inefficiently? If it is 10^6, we might reach that number of neurons very much quicker in evolution somewhere else.

MINSKY: I think 10^6 is enough to create intelligence; if they are all in the right place, then 10^6 is enough to create a rather smart computer in basic operation. But I don't think it would know enough to be very useful. At least our experience is that our programs start to begin to be interesting and versatile when they have got 10^6 bits.

PLATT: You are identifying bits and neurons?

MINSKY: Indeed I am, because I believe that it is risky to attribute too much intelligence to a neuron. I think most probably many of the neurons on the nervous system are organized in groups and in other ways to immunize the thought process from the variability of the neurons. Thought is very precise and reproducible and there are probably all kinds of clever mechanisms to keep the synaptic variations in the neurons from interfering with it. So I give very little credence to theories that would put memory on a molecular level or try to get a lot of information out of a single neuron. We are trying to get money to build a 10^{12} bit computer.

HUBEL: I don't think we know how to think when it comes to bits and neurons. The important thing appears to me to be obvious. If each neuron is shaped like an oak tree and has 100,000 connections on it, then we will be out by many orders of magnitude if we talk about simple numbers of neurons rather than numbers of connections. The second thing is that many neurons are similar, very closely similar to their neighbors, and in some structures you have something that looks like a crystal.

MINSKY: The computer memory looks like a crystal.

HUBEL: That is what I am saying. That is another aspect of the whole thing. If that is the case then we should talk about a very large number of neurons but also a number of kinds of neurons, which is very much smaller. So one really doesn't want to emphasize that which would make a network of 10^{10} or 10^{12} bits. Or accordingly we should talk about the number of connections. We have vastly more connections than 10^{10}. It would be more like perhaps 10^{15} or 10^{16}.

So when you ask how many neurons does it take to make intelligence, I think one isn't necessarily asking a very meaningful question. It may not be the right question.

MORRISON: It is very hard to give meaning to such absolute numbers if you don't know how the components work and how to correct them. But I think the computer guide is for us a very helpful one. There is a very rapid payoff with increase in the number of elements. The function is very steep.

MINSKY: I agree that there are 10^3 or 10^4 more connections than neurons. On the other hand, in the computer 10^6 bits is the minimum to get interesting programs and 10^{12} is much larger than I can imagine necessary for creating intelligence. With a computer one uses these bits very effectively. One can write the program without redundancy. I think your factor of 10^3 or 10^4 is compensated for by the inefficiencies with which neurons are used in the brain. There are related problems that most of the brain neurons aren't in the right place to be used at all. I am just saying that the efficiency of the brain is low.

ORGEL: That sounds like computer chauvinism.

MORRISON: I was going to say that we have some, though few, empirical guides. I was taken by the remarks of Doctor Lee who showed us, I think, that in his view, to go from a highly manipulative monkey to some kind of elaborate social formation was a factor of 5 or so in brain capacity. I suspect this is the kind of very steep function we are dealing with. I think this idea, that intelligence is highly nonlinear in the number of bits, is probably what may turn out to be an important consideration in the time scale of the evolution of intelligence.

The qualitative situation changes when the number of bits changes by rather little, if you are talking about such a fundamental number as the number of interconnections.

Surely the parallel between monkey brain and cattle's brain is pretty close. It is as close as you will find, I imagine, in biology. All their substrates, and so on, are pretty reasonably the same. Yet if we are to believe anything about brain capacity, a rather small change in the volume made a big difference in behavior. I would expect this is a very typical case.

HUBEL: The dolphin has a very large brain.

GOLD: A dolphin has a larger and more complex brain and more neurons than we, and it doesn't seem to do a great deal with it.

SAGAN: Likewise whales. But they may have a very extensive verbal tradition. The lack of manipulative organs like hands may be the limiting step in developing what we can readily recognize as high intelligence. There is a lot of anecdotal data suggesting nevertheless that Cetacea are smart.

PLATT: I would like to ask Lee a question about what he said earlier about the bushmen. You said something to the effect that they have only one hundred words for technological things.

LEE: They have only one hundred words for material items, material culture.

PLATT: But is it not true that they have a tremendous number of words for the environment?

LEE: Of course.

PLATT: They have a name for each tree, for each bird individually, and so on; so it wasn't too clear what conclusion you draw from that.

LEE: My general point was that their intelligence seemed to have evolved beyond their adaptive need and that we can have the coexistence of a highly evolved intelligence with a technological simplicity.

GOLD: But how do you know that their intelligence evolved in the bushmen? How do we know that their ancestors before they were bushmen were not equally intelligent, but living in totally different circumstances from what you now know?

LEE: I don't think we have any basis for that supposition. We have archeological evidence for bushmen going back about 50,000 years and giving indication of something that is in all respects ancestral to bushmen. What sort of alternative ancestor did you have in mind?

GOLD: Fifty thousand years is not such a long time but I imagine that the intelligence—

LEE: But for a small number of material objects, it is. You may be speaking about some hypothetical devolution or fall from grace, but I don't think that is seriously entertained as an hypothesis in anthropology.

SAGAN: I wonder if we can ask Doctor Orgel to spend one minute on what he thinks were the salient issues to the present point of this rather unstructured discussion.

ORGEL: The first point was Minsky's very much to-the-point question that if we adopt Francis Crick's very critical view, what should we do? To which Francis replied, "You can only look where the light is; it is no good looking anywhere else." To which Carl Sagan in return replied that really you do have alternative strategies. It does seem to me to point to one thing, that before we are through we really have to deal with this question because we can't go away in a totally agnostic mood saying that really it is not possible to do anything because we don't know anything.

I thought the discussion on the evolution of the eye interesting but not conclusive, because our experts don't seem to reach an agreement as to whether the degree of similarity between the eyes was surprising or not. The next thing that I enjoyed was Tommy Gold's attack on species chauvinism and mainly his claim of something like an ant that got together and that might look at us and feel that we had a rather poor way of doing things.

The only other note I have of anything which I happened not to be so involved in that I had time to write it down was Minsky in favor of humanoids, which seemed to be an interesting argument, both the Asimov argument, and also his later argument about it being more important to be able to use knowledge than to originate knowledge.

Let me say, in case anyone feels slighted, that if I haven't put them down it is almost surely because I was listening with too great an interest.

PLATT: As several have suggested at this conference, we should generalize equation (1) to include a large number of f factors, or probabilities of various sequential steps in evolution up to the level of extraterrestrial communication. However, the effect of this increase in the number of f factors in reducing the total probability and the total number of ETI's, N, is not as great as McNeill and others have supposed. Table 1 shows the state of our knowledge of these probabilities for some 25 fairly representative steps of this kind. About half of them appear to be "certain," marked with "C," in the sense that we know them to occur everywhere (chemogeny of small organic molecules), or to be theoretically certain (autocatalytic cyclical buildup in a radiation-flow field), or to have occurred independently in multiple instances (convergent evolution) on Earth.

More precisely, in these C cases, the probability of each successive step f can approach indefinitely close to unity, because of multiple repeated independent attempts, if enough time has elapsed. The step is "mathematically certain" in the exact technical sense that $f = 1 - \delta$ where δ can be made less than any given number in a finite time, ϵ.

In most of the other steps in Table 1, the probability is unknown (marked by a question mark) and needs more careful study; although in my opinion and in the opinions expressed by Sagan, Orgel, Crick, Hubel, Flannery, and Lee, most of these steps may follow almost necessarily from the preceding steps and are therefore "certain," or C, in the same sense. (The downward arrows at the left side of the table suggest how several succeeding steps are almost certainly "implied" by preceding steps.)

Three steps, however, might have been unique (marked in the table with boldface type and with horizontal arrows), and at the present time must be assigned unknown probabilities between 0 and 1 (as Crick and others have emphasized) until we have more theoretical or experimental evidence on how inevitable such steps are. These three steps are (1) the origin of life, (2) the use of fire (in my opinion), and (3) the development of symbolic language (which might be equated with the step to "intelligence"). It is therefore appropriate that two of these, at least, should occur as crucial unknowns in equation (1), as f_l and f_i. (The last probability term, f_c, for communication, is represented by the last two lines of Table 1, "astronomy and science, nuclear power and space," which I regard as a "certainty" following the preceding steps, or C, in the sense used above.)

I think a little more work on breaking down these crucial steps into substeps might enable us to show theoretically that they are also "certain." For example, the autocatalytic cyclical buildup of larger molecules in a radiation-flow field was emphasized several years ago by Prof. Antoine Zahlan of the American University in Beirut. One can make a "directed graph" of all possible molecular species, with the vectors of the graph showing the flow-rate of conversion of one species to any other in a given radiation field. (If the radiation is turned off, the rates go to zero because of microscopic equilibrium—vectors equal and opposite—giving a simple Boltzmann distribution of abundances for that temperature.) On this graph, there will be many cyclical loops, A-B-C-D-A-B- . . . , driven by the radiation, and

Table 1. Probabilities of Various Steps in Evolution

Implications	"Certain"[a]	Probability unknown[b]	May be low?[b] (unique?) (crucial?)
↓ chemogeny in radiation field	C		
↓ autocatalytic cycles (Zahlan)[e]	C		
origin of life			←
protein/nucleic acids		?	
↓ first cells		?	
multicellular organisms		?	
(photosynthesis → O_2) (may not be necessary for evolution?)		?	
animals		?	
nervous network/ chordates	C		
↓ eyes	C		
land animals		?	
learning-nervous system	C		
social animals	C		
communication-signals	C		
(mammals) (May not be necessary for evolution?)			
(families) (May not be necessary for evolution?)			
hand-eye coordination	C		
tools	C		
curiosity	C		

Table 1 (*Continued*)

Implications	"Certain"[a]	Probability unknown[b]	May be low?[b] (unique?) (crucial?)
fire (I believe this is necessary for speech and technology)[d]		?	←
language—symbols[e]			←
villages (timed by the ice age?)		?	
technology	C		
cities	C		
astronomy and science	C		
nuclear power and space	C		

[a] "Certain" is assumed if there is evolutionary convergence of several independent instances. This means $f = 1$; or rather $f = 1 - \delta$ where "δ can be made less than any given number in a finite time, ϵ" so that achieving such steps are not matters of "probability" but of *time* (assumed $\ll 10^9$ years).
[b] Problems needing special study.
[c] Can autocatalytic cycles guarantee origin of life? (Or are they equivalent?)
[d] Can *fire*, by *lengthening the day*, plus *social animals/families* guarantee speech?
[e] What is probability of *fire* and/or *symbolic language*?

among these loops there will almost certainly be many auto-catalytic loops of buildup and splitting, A-B-C-D-2A-2B- . . . . No matter how low the probabilities of such buildups may be, they will certainly lead to a buildup with time of all the molecular species involved, A,B,C,D, . . . , and of all their breakdown products, and so to the sequential buildup of all more complex autocatalytic loops for their precursor molecules can then lead to "natural selection" for more successful autocatalysis in various environments, as a function of time.

Many people might regard this combination of "autocatalysis-plus-natural-selection" as being equivalent to the "origin of life";

but in any case, it seems to make the origin of more complex "living systems" almost inevitable, as suggested in Table 1. I believe this conclusion follows inescapably as soon as we take not molecules but autocatalytic loops or *systems,* as the units on which natural selection operates.

Likewise in the case of the development of symbolic language, the probabilities may become much higher as soon as we consider substeps and precursor steps such as, perhaps, the use of fire. I am not sure of the probability of the use of fire, because of its danger and repellence to many animals, but I think many anthropologists would regard it as "certain" (in the sense above), given a population of tool-using land animals with curiosity in an oxidizing atmosphere.

But once fire is mastered, it seems to me this makes almost certain the transformation of animal communication signals into symbols and language. The reason is that fire extends the day, in time into the night or in space into caves and dark places. As Lee mentioned, this enforces leisure, and the repertoire of imitation and communication signals, cut off from their daytime referents of hunting, or gathering, or fleeing, can only continue to be emitted as symbols of things remote in space and time, giving rise to drama, ritual, history, poetry, myth, and science.

The overall result of these considerations is shown in Figure 10. It is clear that the probabilities f in equation (1) must be considered as functions of time. This immediately illuminates some of the controversies at this conference, such as those between Crick, McNeill, and Sagan. As shown at the top of Figure 10, once the precursor steps are past, each of the f's grows in time, in a given range of planetary environments, and many of them may approach 1.00, or "certainty," in times short compared with cosmic times. McNeill emphasizes, correctly, that the product of a large number of f's less than 1.00 is essentially zero. But it is also true that if they all approach certainty, 0.99, 0.999,

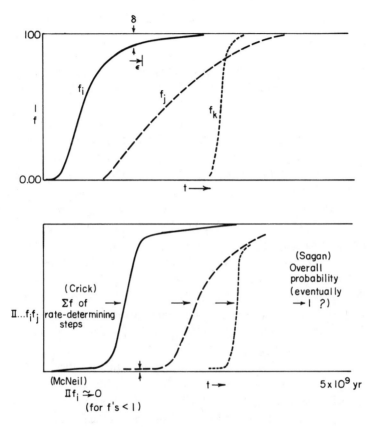

Figure 10. Views of the time development of the constituent probabilities
f in equation (1) for the development of technical civilizations on a given
planet. See text for details.

etc., the product of a large number of them may still rise and
approach unity in a finite time, and Crick emphasized that it
might be more fruitful not to look at the f's, but at the sum of
the times t of the rate-determining steps. On the other hand,
Sagan and Drake (insofar as equation (1) is time-independent)
were less interested in the time-dependencies than in the overall

product of the f's at the end, in cosmic times which are (hopefully) long compared to all the rate-determining steps in the planetary systems of interest. The latter assumption may, of course, be false, and there may be many types of planetary systems which have crucial f's with billion-year time constants or longer.

LEE: I can't speak about autocatalysis but I can say something about the origin of language. I followed Doctor Platt's argument with mounting excitement hoping that he would do something that hasn't been done before, assign a probability to the origin of language, but then he said: "Aha, here is the problem"; so you independently, I guess, have arrived at the conclusion that the anthropological profession has arrived at, that there indeed lies the problem.

So, I don't know if you have taken the argument further than reiterating our ignorance. Language is, indeed, one of the most difficult things that we constantly have to come to grips with. We have not yet made a breakthrough in understanding its origins. But I agree with you that tools, fire, and the rest don't offer us too much difficulty.

OLIVER: I think that what, in fact, happens is that intelligence develops for the various reasons that have been known since Darwin and Wallace, and as intelligence develops a number of these steps automatically takes place because of the large intelligence of the evolved individual or species. For example, I do not hold that the beginning of language is such a remarkable step. I believe that many animals, for example Cetacea and birds, have rudimentary language. The difference between their speech and ours is a question of degree and of abstractness, not of existence or nonexistence; so I think that we really have to remember always that when we view our past, we are looking at a particular way in which man got to where he is, but that there may be alternative paths that converge on a similar destination.

If each of us as an individual were to assess the probabilities of the mating of our ancestors far enough back into the past, each of us would conclude that we, as individuals, were totally impossible.

GINDILIS: Professor Crick has said that we must clearly distinguish between the question of the uncertainty in determining a factor such as f_1 and the question of the value of the factor itself. I agree that the reliability in assessing such factors is indeed very low but I would like to say a few words in defense of the actual value of this factor. I am not proposing any numerical assessment but would merely like to remark on the approach to determining f_1.

Very often the probability of the origin of life is determined as the probability of a random emergence of a thermodynamic fluctuation that can give rise to the formation of a complex system of the type of DNA or a protein molecule, and since such a system is rather complex, we need not be surprised that the probability of such an event is negligibly small.

Professor Sagan noted in an example of such a calculation that the probability of the random assembly of a rather simple protein is 10^{-130}. On the basis of such calculations, it is sometimes inferred that life is a rare, exceptional phenomenon in the universe and that our Earth is simply lucky in this respect. I would like to say that the erroneous character of such an argument is evidence of the fact that such a purely combinatorial approach is inapplicable to the process of the formation of a complex system.

By simply combining the initial elements, one cannot in a reasonable period of time obtain even the simpler systems which exist in nature, let alone a protein molecule. One can raise the related question of a purely random formation of a sodium chloride crystal which consists only of sodium and chlorine, but where the number of atoms in the crystal is great. We would take 2 to a power equal to the number of atoms, and obtain for the probability of a crystal of salt a quantity as negligible as the figures that were written on the blackboard by Professor Sagan.

Of course, the objection is that in this case physical laws come into play and not merely chance. But it seems to me that the same physical/chemical laws operate in the formation of complex compounds, compounds that are the precursors of life. Complex biological systems may evolve in such a way that at each stage there are certain intermediate subsystems that, by virtue of their structural features, seem to rule out the possibility of the formation at subsequent stages of the process of many of the a priori possible combinations of simpler elements. At every stage of the process there thus arise only certain permissible combinations, and this radically alters the picture.

I would like to explain this idea by citing the well-known example of the system of language. Written language has as its elements the letters of the alphabet. From those letters, by certain regularities, we obtain words. From words we build up sentences and here, in turn, certain laws again govern the process. This leads to the result that a vast array of the a priori permissible combinations of letters are never realized in a meaningful text.

It is well known that there is a theorem in information theory that the number of such senseless sequences is far greater than the number of meaningful sequences, but their total probability may be as small as you like.

I think that the same is true in the process of the formation of complex material systems. In this sense chemical evolution too is like a narration. One passage of its evolutionary narrative complete, nature predetermines several subsequent "letters," "words," and "sentences." But what nature will have written millions of years hence, we do not know. I visualize chemical evolution as an ergodic process which can be described as possessing a certain redundancy, and the greater that redundancy, the closer does the process come to a thermodynamic one.

But we know that for any random process, there is a characteristic time after which that process inevitably takes place. This is all the more so in the case of a quasi-dynamic process, although here the characteristic time must be determined with due consideration for the quasi-dynamic connections I referred to; and that must entail a reduction in the characteristic time. In the final analysis it must increase the probability of the realization of the process as a whole.

If that characteristic time in the process of the formation of life, the emergence of life, is less than the time for the existence of the planet, life on such a planet must arise inevitably. Therefore, the factor f_1 may be interpreted as the probability of the characteristic time for the emergence of life being less than the time of the existence of the planet.

We know that on the planet Earth this condition was observed. However, that characteristic time for the earth, as has been stated repeatedly and as was pointed out in our discussions by Professor Crick, may be of the same order as the age of the earth. The slight difference in physical conditions on other planets could increase the characteristic time by 1 or 2 orders of magnitude. Then for the emergence of life on some of these planets there would have been a need for a time greater than that available.

A consideration of this is important since it may reduce f_1 substantially. But if the reliability in determining that factor is small, things are no better with regard to other factors.

Unfortunately, it seems to me that at the present level of our knowledge we have no possibility of even roughly estimating the number of civilizations in the universe, and that being so, I look upon the proposition concerning the existence of civilizations as a hypothesis. This hypothesis seems to me to be very probable, but this does not make it any less a hypothesis. It is still a hypothesis, and in this respect little has changed compared with what we had fifty, one hundred, or a thousand years ago. Indeed, the hypothesis of the plurality of inhabited worlds is just about as old as ancient Hindu philosophy.

One thing however has changed fundamentally. The fact is that now for the first time over the course of the development of science, there are means of verifying this hypothesis. First of all, I refer here to the fact that modern radio facilities make it possible to register signals sent to correspondents at interstellar distances. Of course, the possibility of detection depends on the power at the command of the sender, how far away he is, and upon many other factors, to be discussed later in this symposium. Here I would merely say that what is important is that we are able to detect these signals, although under certain conditions. But this detectability in itself radically alters the picture since it makes it possible to ask that there be experimental investigations in this field.

Such investigations can at the outset be based on the achievements of astronomy, radio astronomy, cybernetics, mathematical theory of communications, and so on. They have something to fall back on. For this reason, it seems to me that we now have the technical facilities and the methods for starting the work of detection. Given these conditions, it seems to me it would be wrong to require independent evidence confirming the existence of extraterrestrial civilizations before we begin this systematic search.

ORGEL: When Doctor Crick and I say that the probability of the origin of life cannot be estimated with any confidence, we do not base our opinion on the supposition that life originated as a single combinatorial accident. It is our opinion, as a result of experimental and theoretical work on the subject, that our science has not yet progressed to the point where a meaningful estimate can be made.

However, although it may be impossible to determine the probability of an event, it may yet be possible to say that that probability is increased by a series of observations. Therefore, I would disagree somewhat with the previous speaker when he said that nothing that had happened in the last fifty years would change our opinion of the probability that life has arisen. It cannot be denied that Stanley Miller's experiments, which showed for the first time that organic materials of the type that exists in terrestrial organisms can be made under remarkable and surprisingly simple conditions in the laboratory, have increased the probability—have increased our estimate of the probability—that life could originate spontaneously; but, unfortunately, have not yet increased it to a point where we can assign a high value with any confidence.

MUKHIN: I do not quite understand how we can estimate f_i in equation (1), when we cannot choose any rational approach for assessing f_i. I think it is correct to say that a reliable estimate cannot be given. Perhaps we must accept these probabilities equal to 1, since we are all here and this is proof which cannot be ignored. Many other estimates will be completely arbitrary. But, others say, so long as we do not know all the stages in the process of the origin of life on Earth, we cannot assess the probability of that event.

It seems to me this is not quite legitimate. In my opinion, it is best to proceed on the assumption that in the universe there can be no unique objects.

SAGAN: We are faced, as I have said several times before, with very difficult problems of extrapolating from, in some cases, only one example and in the case of L, from no examples at all. When we make estimates we cannot pretend that these values are reliable. There is no sense of statistical probability which we are applying to the later factors in equation (1). We are making subjective probabilities in the sense of Professor Fine (Appendix

A). Their *only* value is in assessing how much effort, time, and money we are willing to devote to the problem.

The history of science is replete with a kind of grand principle very similar to what Doctor Mukhin mentioned, saying that what we have on Earth is in no special way unique, a view associated sometimes with the name of Copernicus. I would like to mention one case (there are many of them) in which this simple assumption allowed quantitatively correct estimates. This is a calculation performed by Huygens and independently in a slightly different way by Newton. Huygens argued that the sun was a star, that it was in no significant respect different from other stars. He knew of the inverse square law of the propagation of light and therefore asked himself how far away would the sun have to be in order to be as bright as, for example, the star Sirius, which he assumed to have the same intrinsic luminosity as the sun.

He then made a set of holes in a brass plate, holes of different apertures, and then held the plate before the sun, making an estimate of which hole was as bright as he remembered Sirius to be the night before—not a very accurate photometer; but the remarkable thing is that this calculation gave a distance to Sirius of something like half a light year. And if he had known that Sirius was intrinsically much brighter than the sun, he would have made an even more accurate calculation.

This is one example—there are many others—in which it is possible to do semiquantitative estimates from what I call the assumption of mediocrity, that is, that what is on Earth is pretty characteristic of what is elsewhere. This is not a rigorous argument but I believe it is a kind of guide.

DEBAI: I want to speak of some factors that have not been mentioned yet. I am speaking of the influence of the moon. Life arose in the liquid phase. A technological evolution is possible on solid ground and on the earth this was connected with lunar tides, in the course of which marine animals learned to breathe.

If this is so, it imposes a severe limitation on a repetition of such a world, in that the absence of a suitable satellite, at a suitable distance, would reduce by a large factor the probability of the evolution of large animals.

My second remark concerns the criteria concerning the artificial and the natural. I completely agree that there are no unique objects in the universe; this will probably enable us to distinguish between natural and artificial signals. The fact is that natural objects obey selection laws whereas the human imagination suffers from no such limitations. Therefore, the similarity of one phenomenon to another displays the clear signs of its natural character.

I will refer to two well-known examples of this: first, the discovery of CTA 102, the well-known radio source. When it was the only one, it was promptly declared to be artificial, but when many such sources appeared, the idea of artificiality was abandoned. The second example concerns the pulsars. The first pulsar was not declared to be natural or artificial and the observers very cleverly waited for the discovery of other such objects. The existence of a large number of such objects immediately made it plain that they were not artificial.

So, as you see, nature is uniform and intelligence is diverse.

AMBARTSUMIAN: How many years do we need to wait for the discovery of a second object for our problem?

GOLD: I would like to say two things about the tides. One is that I believe the existence of the tides makes the in-between situation harder rather than easier. It would be harder for an animal to adapt itself to land from water if it had to survive the intermediate circumstances of the washing of the tides.

Second, the absence of a moon would still leave solar tides a third as high as the present ones in any case. So I do not think there is any great difference to be expected in real life between a circumstance with a moon or without a moon, so far as adapting from water and coastal land is concerned.

TOWNES: I find it very difficult to make the kind of overall absolute statements that are attractive to make. For example, in the case of the moon, I think it is quite unclear as to the advantages or disadvantages from an evolutionary point of view. There is simply, for example, the counterargument that difficult conditions are in nature just those which lend themselves to rather rapid evolution, and it is just in the very even, completely uniform, regions where evolution moves slowly. I don't present this as an overwhelming argument that the moon is good for evolution, but, rather, as an illustration of how difficult it is to make certain statements.

I would then like to discuss the point raised by Professor Sagan. Professor Sagan was appealing to the principle of mediocrity. I find it difficult to make very firm conclusions from such analogies. The principle of mediocrity, for example—at least I believe this is in the same spirit—led human beings to think that life was being produced continuously all around us, that frogs were spontaneously generated from the mud every year. This was disproved through long and hard work by Pasteur, while it seemed quite contrary to the general human instinct of the naturalness of life.

Life, in fact, at least on some relevant scale, is quite unique and not continuously generated. It seems to me the same kind of spirit of mediocrity is that which has led people to believe that everything that we have done in the past can't be improved. And yet discovery itself is a unique event in the life of mankind. It is only in the belief that there can be changes or new situations that we are willing to move away from that tradition.

Now, I must caution that these, of course, are simply analogies and I disbelieve in such analogies, but in the same sense I disbelieve in the analogy which Professor Sagan gives as being fairly convincing.

CRICK: What Professor Townes has essentially said is what I

was going to say, but he said it so politely that I would like to say it again. You cannot take a hypothesis and then from an estimate of its probability compare it with a second hypothesis unless there is some structural analogy or unless you know the probability of the second hypothesis.

SAGAN: Professor Townes was perhaps too polite but the words he used were that one could not in this way reach "very firm conclusions" and that one should not assume that these were "fairly convincing arguments." My goodness, we certainly all agree with that. No one is maintaining that there is any degree of rigor in such arguments. I would turn Townes' spontaneous generation argument around, and say that since people are not spontaneously generated the assumption of mediocrity implies that frogs aren't either.

I think the problem is that we are used to making probability estimates in very different contexts. Let me try to say what I think is the function of what we are doing. The reason we want to estimate these values of f is to decide whether it is out of the question to search for extraterrestrial intelligence. We have the technological capability to do it. The question is, shall we proceed?

If it turns out that there is some rigorous argument to exclude extraterrestrial intelligence, a convincing demonstration of a small value of N, then a search would not be a useful allocation of resources. If, on the other hand, we cannot exclude it, then I think it is very likely that we are going to proceed to try and find it.

From some of the discussion that I have heard, it sounds as if the fact that there is life and intelligence on Earth is totally irrelevant. I am very puzzled by this. I would have thought it would have had some at least minimal relevance to the problem. It is the only case where we happen to have data. It is not enough to make preliminary statistical conclusions on, but it surely is of some relevance. If one case is not relevant I wonder how many cases are necessary before we get relevance.

CRICK: Two.

SAGAN: Two cases are convincing, I maintain. One case is not irrelevant. But your answer again emphasizes the importance of searches for simple forms of life on Mars, Jupiter, and elsewhere in the solar system with sterilized spacecraft.

I stress again that, in the context of this discussion, no one pretends any even moderately high reliability to these estimates. The question is, Is there some argument to exclude extraterrestrial intelligence, in which case it would be a waste of time to look for it? If we cannot exclude ETI and if a search uses instrumentation which in any case would be useful for non-ETI-related astronomical activities, then sooner or later such searches will be initiated.

MORRISON: If there is anything more to be said about these probability arguments, it can be said by the probability theorists; it is a complicated problem and I think they can handle it themselves. It seems to me that eloquent proponents of both positions have had their say.

THE LIFETIMES OF TECHNICAL CIVILIZATIONS

L

SHKLOVSKY: I have some difficulty in defining for myself the subject of our symposium. It seems that we are discussing a new science, a very important science and a very interesting one. But all the natural sciences, as you know, rely upon observations and experiments. We have nothing of the kind here. What we are postulating here is that apart from the highly organized civilization we know here on Earth, there are other civilizations elsewhere in the universe.

Other sciences—physics, for example—have such fundamental postulates. I need not go into examples, but those postulates, as you know, are generalizations of experimental findings. In our case, the postulate—the postulate of the plurality of inhabited worlds—rests on logic alone. In this sense, the subject of our symposium is similar to geometry which is also based on logical postulates, but I think that even the most faithful among us are not as convinced of this plurality as the experts in geometry have always been convinced of the fact that parallel lines do not meet.

Equation (1) contains a number of factors. In our previous

discussions we have concerned ourselves with all factors except
L. Our discussion revealed quite a number of certainties concerning them. Concerning one of these factors, I would like to
say, by way of illustration (although I certainly am not an
anthropologist and I speak as a layman), that I once came across
a hypothesis in the literature that the development of a species
of prehuman primates into Homo sapiens was the result of the
fact that our ancestors were suffering from a particularly un-
pleasant parasite.

I have the feeling that the subject matter of our symposium may
not yet be really called a science in the strict sense of the word,
although this observation does not detract in the least from the
importance of what we are discussing.

This is a multifaceted problem which rests on a hypothesis that
seems quite tenable, and this hypothesis rests on the assumption
that among the 10^{22} stars in the observable universe there are
some with planetary systems, among which there are planets
capable of supporting intelligent life.

The fact that a highly developed civilization can influence the
planet which it inhabits may be illustrated by a rather simple
example. It is an example I gave about ten years ago and which
seems to have gained some popularity. It is the fact that, thanks
to the development of television, the brightness temperature of
our planet in the meter wavelength range has, on the average, become something of the order of $10^{8}°K$, and this makes our planet
second in brightness only to the sun in our solar system. In some
wavelengths the effect may be even greater; in some narrow bands
used in radar studies of the planets, the order of magnitude of
such powerful emissions is far greater than the emission of the
sun.* However, this is an effect which would be extremely difficult

* Editor's note: Direct observations, showing radio brightness temperatures of 4000°K at 73 to 440 megahertz from only single towns in Illinois
have been reported by G. W. Swenson and W. W. Cochran, Radio Noise
from Towns: Measured from an Airplane, *Science* 181 (1973): 543–544.

to detect from a distance such as that to the nearest star, because we are dealing with a quasi-isotropic emission rather than a directed one. At 10 parsecs, the flux of the earth's emission due to television would be about 10^{-37} watts per square meter per megahertz.

We can, of course, visualize a civilization at approximately our level, but commanding even greater power. However, I think that a civilization of this type could not make its existence known in the galaxy by such means, but of course I am speaking of isotropic emissions and not the beaming of signals which will be discussed subsequently.

In recent years, the idea has become current that there may be civilizations commanding much greater power resources, civilizations of what we call Type II and Type III. If I am speaking not of beamed signals but simply of eavesdropping on isotropic emission, this could apply only to civilizations of such high classes as use energy of the order 10^{30} or more ergs per second, 10 billion times greater than the energy at our command—and I use "billion" in the American sense.

But the conditions and the strategy of such a civilization must, of course, differ radically from ours. It is my own personal view that the advance from a civilization of Type I to a civilization of Type II implies the appearance of a new factor in equation (1), and this factor may be negligibly small. I think that such very advanced civilizations must not be biological but, rather, computer-devised and spread out over enormous areas. It is even now becoming clear that the existence of biological systems in environments which command such enormous energy resources would be extremely difficult.

In such a situation, we are faced with entirely new prospects. I can mention them but briefly. For example, the radiation hazards that can be so fatal to us would be irrelevant here, and such civilizations could emit hard radiation at very short wavelengths.

Common to all the civilizations must be the re-emission of as much energy as they consume. This principle is basic, because otherwise the temperature of such civilizations would become incredibly high—impermissibly high. I would like to hear the view of specialists on the following question: Need such a system necessarily emit only in accordance with Planck's law, or may it emit energy in some other way?

Further, such civilizations need not necessarily confine themselves to using the energy of their central star. For example, using the energy of such a large planet as, say, Jupiter would make it possible to produce energy of 10^{35} ergs per second—more than the sun produces—for hundreds of millions of years. Perhaps for civilizations so far advanced the harnessing of the energy of the central star may be too primitive and naïve an approach.

I would also like to underscore that the development of such advanced cybernetic civilizations may be described as a logical abiological development of life as we know it. It may be that what we call civilization is merely an intermediate stage on the road to a far more advanced civilization, an intermediate and unstable step, moreover.

Finally, I would like to mention what I consider personally to be a very important matter. When we consider emission energies as enormous as 10^{30} or 10^{35} ergs per second, considering that this may be in any arbitrarily chosen wave band and, moreover, a product of the biological activity of such civilizations, this prompts a question: How are we to distinguish such a signal from a natural signal? Although seemingly elementary, this is really a very formidable problem. I could give quite a few examples from recent astronomical observations or developments which suggest that at first we believed we had been observing what we call a cosmic miracle, but that miracle, upon closer scrutiny, proved to be a perfectly natural phenomenon. Therefore, the criteria for the artificial nature of such signals becomes one of our major problems, and I hope they will be discussed here.

I would like to conclude by saying that in investigations of this matter, we must follow the legal principle of presuming signals natural until proved otherwise.

PLATT: Can L [in equation (1)] be determined by CETI? Advanced civilizations may be like the parents who do not talk to the baby until the baby wakes up. In that case, we can make no estimate. Here on Earth, we are in the midst of a great watershed, a world transformation. In the last one hundred years, we have increased our speed of communication by a factor of 10^7, our speed of travel by 10^2, our energy resources by 10^3, our weapons by 10^6, our data processing speeds by 10^6.

Some of these are sudden transformations, such as radio. Some are more gradual. Some, like population, are rather slow but altogether many of them are approaching certain natural limits. We cannot communicate faster than the speed of light or travel around the earth faster than orbital speed, or increase our energy beyond the thermal pollution which now threatens Los Angeles; and on the weapons side you cannot be deader than dead.

Data processing may increase by another factor of 100, but altogether we are in a transition period on a scale such as no society has ever encountered. None of our social organizations are prepared to deal with changes on such a scale. The result is that we may oscillate, or we may destroy ourselves, or we may reach a high level steady state.

It will be a new form of society, totally different from anything that has ever existed in the world before, as radically different as a new species, if we survive.

The catastrophes include nuclear destruction, pollution, ecological disruption, overpopulation, and exhaustion of natural resources. Over half of our critical minerals will be exhausted in twenty years. The half-life estimated by Leo Szilard for nuclear annihilation was ten to twenty years, if we continued to have more nuclear weapons every year which would lead to nuclear confrontations and possible accidents.

What is needed, I think, are two things which determine whether we can go to a higher level of organization or whether we wash ourselves out. The first is an understanding of conflict resolution. As in the books by John Burton on *Communication and Conflict* and the books by Anatole Rapaport on nonzero sum game theory, our most dangerous problems, many of them, are in the area of nonzero sum games. These are cases where individually rational behavior nevertheless produces collective destruction. It is a narrow rationale. Rapaport's books are entitled, *Two-Person Game Theory* and *Prisoners' Dilemma,* and I might add one called *The Big Two.* You can imagine who the "Big Two" are.

The second need is systems analysis of these interacting problems because the interaction makes them all more dangerous. A start has been made on systems analysis by Jay Forrester of the Massachusetts Institute of Technology in his book, *World Dynamics.* He does computer simulations of population, food supply, pollution, consumption of natural resources, capital investment, and so on. Each of these has an effect on the other.

He assumes coupling constants that may be wrong, but the result is that he can simulate further into the future how pollution will rise, how food may fall behind population, and how natural resources may be used more violently. The result may be that the population has sudden catastrophes. The time and the scale of these catastrophes depend on what policies we make, and a typical time for his population catastrophe is the year 2020. This work may be wrong, but it is a first step—it is important for others to do it better than this first step.

Now let me turn to nonlocal intelligence. I think that the lifetimes of extraterrestrial societies may involve a race between cooperation and competition. Cooperation is needed to develop language, symbolic communication between individuals, while competition may be necessary for fast technology development, as Doctor Gold suggests.

I can imagine three types of societies: first, one which has solved the social organization problem before high technology. This might be a society of very docile animals or very dependent animals, or it might be a society of termites.

The second type would be a society which also solves social organization by an accident, such as a nuclear dictator as, for example, Hitler. If he had the atomic bomb, he would have wiped out all the resisters in the world, but there might have been a society continuing which had high technology.

The third case is ours. This is where the social organization or conflict problem is not solved before technology, in which case I think we now have to make a probabilistic discussion. I think this problem is like a threshold problem and not like a half-life problem.

Many societies may be destroyed, may destroy themselves before they solve this problem, but some fraction of them may solve the problem and survive until they meet the next problem, and so one of these problems might be exhaustion of resources before a technological takeoff. One problem might be conflict resolution before they wipe themselves out. One problem might be a loss of interest. They might turn to internal religion, like Zen Buddhism, emphasizing the here and now. I think each of these thresholds will cause a loss of some ETI's—perhaps through drugs, perhaps through genetic degeneration that we do not understand. Nevertheless, a small number of societies—who knows, 1 percent?—might find out how to solve these successive problems and survive—survive 10 million years, 10 billion years?

I think we do not know the possibilities of any of these situations. I think we can say something about the immediate problems on Earth but I think it will be very hard for us to foresee what kinds of problems the new society will encounter.

My conclusion is that 50 years $< L <$ 1 million years. But L might be 5 years.

BRAUDE: Doctor Platt, in speaking about the end of our civilization and in showing this by a curve ending at the year 2020—what was the stability of that curve? Will the parameters change much, say, to make this further away in time, or, say, to make this curve turn upwards?

PLATT: Society is not to be predicted in the way of physics because it is a cybernetic system, like driving an automobile, and the result is that this catastrophe will happen if we do nothing at the present time. Forrester has computed at least twenty sets of curves making different assumptions—about our control of pollution, about our control of the birthrate, about our capital investment in agriculture, and these lead to catastrophes at different times and at different scales; he even has one set of curves which leads to a steady state, which looks very pleasant by comparison.

BRAUDE: What about an insignificant change in the parameters upon which the curves are based? Will the curves remain steady in that case?

PLATT: One of the interesting outputs from the Forrester study is that some parameters are relatively unimportant. For example, the absolute level of population makes little difference whether it is 4 billion or 6 billion, or only perhaps a five-year difference in the time of the catastrophe.

On the other hand, there are coupling constants which are probably sensitive to one-fifth of one percent, and it is this kind of analysis which will identify the sensitive parameters that we need to work on.

STENT: Like Doctor Platt, I also think that the human condition is about to undergo a radical change. The new condition I refer to as the Golden Age. In that Golden Age the problems which arose when, as Doctor Lee pointed out, man left the condition of the happy hunter and became civilized, are about to find their resolution. Thus I am an optimist.

I have set these ideas down in a little book entitled *The Coming of the Golden Age*, in which I tried to demonstrate that the most characteristic feature of this new human condition is the disappearance of creativity from the human scene. The past history and the present state of the two most important manifestations of creativity, namely, the arts and the sciences, show, so I allege, that these two activities are now reaching their happy end. I developed this argument over what I considered to be a rather broad front, and here I can hardly give more than a caricature of my views.

The single part of my argument that I want to discuss here is one which happens to be relevant to the estimation of the factor L in equation (1) and is of a psychological kind. For it is my belief that progress, the fruit of creativity, embodies within it an intrinsic psychological contradiction. By progress I mean precisely the kind of progress with which we are here concerned, namely the gaining of greater dominion over nature. Hence at a time t_2, there has been progress if the actual dominion over nature is greater than that of a previous time t_1. Only practical parameters such as energy expended per head, or speed of travel, or gross national product, are relevant to this notion of progress. It is, therefore, an essentially amoral concept and has nothing to do with affective aspects such as human happiness.

My point of departure for this argument is the recognition that there exists a human character trait that causes people to want to exercise power over nature, that causes them to want to manipulate the environment. This trait was called "will to power" by Nietzsche, although he used it partly as a metaphysical concept, as the very essence of life itself. I merely consider the will to power a psychological attribute which arises during infancy, like any other trait, as the result of a dialectic between the developing brain and the conditions of the environment. I believe that for the evocation of the will to power in this nature-nurture dialectic during infancy, an essential ingredient in the environment is economic insecurity. That is to say, the child, as soon as it is old enough to become aware of the outer situation, is made to feel that unless he exerts himself against nature and tries to dominate it, he will not survive economically —that is, he will starve. It is the will to power that caused man to invent civilization ten thousand years ago.

But with the rise of civilization, as has been outlined for us by Lee and Flannery, and particularly with the appearance of the first economic surplus, it became possible to sublimate this will to power to higher spheres beyond the concern for daily survival, beyond the strife for the next meal. Eventually there appeared an extreme form of the sublimated will to power, which had come to be almost completely divorced from economic activity, a psychotype which some German philosophers of the nineteenth and twentieth centuries have referred to as Faustian man. Faustian man is the embodiment of the extreme will to power. His view of the world is one of constant struggle. And since he sees in struggle the very essence of his existence, Faustian man is never satisfied. Ironically he is that ideal type whom the economists of the nineteenth century had in mind with their idea of the economic man. And Faustian man became the mainspring of progress. That is, being never satisfied, he was always struggling, always pushing ahead, always trying to gain greater dominion over nature, even if his own personal situation was, in fact, economically secure.

And now we reach the contradiction: as the efforts of Faustian man succeed, as progress unfolds, as ever greater general economic security is achieved, an infantile environment arises in which the evocation of the will to power in the child is no longer the normal ontogenetic outcome of childhood. That is to say, adults now appear in whom the will to power is no longer a prominent character trait. One of the first recognizable cases of such a decline of the will to power occurred in seventh century T'ang China, where a degree of economic security, law and order, and cultural advance—both artistic and technological—had been achieved which exceeded anything seen before on the face of the earth. At that time an essentially anti-Faustian, philosophical outlook became popular in the Middle Kingdom, which sought harmony with rather than dominance over nature. I find it characteristic that the first massive acceptance of this idea occurred in the first general ambiance of economic security.

In the cultural backwater of Europe, this development did not occur until the Industrial Revolution, and it is, I think, no coincidence that Romanticism, or the Western (and mainly German) apotheosis of Faustian man, first went into decline in the middle of the nineteenth century. Since then the waning of the will to power has proceeded apace, and with it the gradual exit of Faustian man from the stage of history, until, after some ups and downs, in the so-called affluent society, especially in metropolitan America, this characterological evolution has become a very pronounced feature of everyday life.

And so the self-limiting contradiction inherent in progress that I have tried to set forth here is that the unfolding of progress works against the perpetuation of the psychological force needed to drive it.

One further historical example illuminating the consequences of this contradiction that I want to set before you is the case of Polynesia. Here general economic security was obtained not because of an advancing technology, but simply because an adventurous and courageous seafaring people managed to find a paradise on Earth, where abundant nature then provided for everybody what high level technology is at last about to provide, or has already provided, for many people elsewhere. By the time the first Europeans intruded upon that scene, the Polynesian personality had undergone a profound change from that of their ancestral explorer-settlers, and Faustian man had vanished from the South Seas. I believe that the state of society characteristic of Polynesia is one that we can look forward to in the industrial nations.

So, then, if this argument concerning the psychological self-limitation of progress has any merit, and if, furthermore, it is of cosmic validity (which it may not be), then it would follow that the factor L is quite small, i.e., that the lifetime of technologically competent societies eligible for CETI is quite short. For it is probable that their advanced technology would provide also for other intelligent beings the same economic security that ours is providing for us. And if their brains resembled ours (which they may not, of course) then all those advanced societies would make up a galactic Polynesian archipelago whose tenants are mostly concerned about their inner life and not anxious to communicate with beings in other planetary systems, and particularly not with us.

PLATT: Why shouldn't you have a feedback situation with one generation turned on, the next generation turned off and the next generation turned on again? It sounds like a stabilized situation, to me.

STENT: I consider the question of oscillations not very critical for my argument. For instance, to take the most extreme possibility, if there were a nuclear holocaust here in the near future and just a few people survive, then of course there would be a dramatic drop in the level of civilization and a disappearance of the affluent society. But depending upon how many people, how much knowledge, and how much equipment survives, the survivors would climb back up, probably in considerably less time than the 10,000 years it took the first time to reach economic security.

As far as other possible causes for diminished affluence are concerned, it seems to me that the negative feedback of economic security in the ambiance of the infantile environment on the development of the will to power would require rather severe reductions in the standard of living between generations in order to allow for the will to power to reestablish its former hegemony; but in any case, my main point is that there will be some kind of non-Faustian steady state at any high level of technology, which is also the point that Doctor Platt makes.

BURKE: On a note of optimism, I would address the soft Apocalypse that Doctor Stent has raised. I think it is generally agreed that the threshold events in technology hinge on the activities of only a few people and most societies tolerate a rather large range of behavior among its members, although, of course, that is not always true. For example, our present society consists mostly of people producing useful goods, but it also allows a few academics to discuss abstruse subjects. Therefore, I do not see why a society consisting largely of people who do nothing should not allow us the same pleasures and, indeed, allow us the tools with which to do what we will.

STENT: I am sorry that I apparently did not make myself more clear. My argument did not touch on the question of disallowing individual freedoms or the tools for any activity in the future, nor did I forecast that creative persons will soon be forbidden to exert their creativity. My point was that the new conditions of the infantile environment no longer favor the development of personality types from which in later life the few creative individuals are recruited on whose activities the technological threshold events hinge. Creativity, or the intensity of the will to power, is distributed in any population and, in line with Doctor Burke's comment, the few highly creative individuals form the narrow tip of a steep pyramid with a broad base. Thus my contention is that the distribution of creativity will alter radically, so that it will resemble a squat pyramid with a greatly diminished distance between the tip and the base. This will happen as a consequence of the altered childhood environments, and not as a consequence of any suppressive acts by which the majority suppress the minority.

BURKE: Will there be no distribution of such environments?

MORRISON: Will no one make mistakes?

STENT: Yes. But I think the distribution will shift only slightly.

MINSKY: I agree very much with the spirit of Doctor Shklovsky's remarks. I think within eighty to one hundred years we will have the capability of building machines of enormous intelligence. I think that may be a factor not in the models of either Doctor Stent or Doctor Forrester; that is, one of the new features of the situation is intelligence, which is probably going to increase by a factor of 10^4 or 10^5 in that same period.

There are a number of advantages, as Doctor Shklovsky indicated, to becoming mechanical. One has a sentimental attachment to one's biological shell and most people who are culturally conservative will want to stay in their bodies because of various well-known advantages. But there will be others who will be intrigued by the possibility of a few improvements such as immortality, colossal intelligence, the ability to experience a wider range of abstract and concrete phenomena that are beyond the reach of humans. I think each race that reaches the kind of critical technological point that we are at now will have to make a decision which may very well be a decision between the horrifying Golden Age of Doctor Stent based on an incredibly primitive psychological theory of power, as opposed to any number of other psychological theories that one could invent.

The possibility that I see and that Doctor Shklovsky sees is that a technological society can convert itself into a species of small, powerful creatures that are perhaps durable enough for their own interstellar travel and technological enough to harness the power of a sun or a large planet.

Now I find that ten minutes is too small a time to explain how this will come about and, besides, I don't exactly know. In a way, my position is a little bit like that of an exobiologist, namely, I am in the field of artificial intelligence and often an audience reacts the same way to an exobiologist [cf. Appendix B], namely, it says, "You are studying a field that does not exist." But in the last fifteen years we have made a factor of 10^6 improvement in the intelligence of our computers, and I will simply have to play poker with you and say that I have here in my hand some evidence that artificial intelligence exists: it is the Massachusetts Institute of Technology doctoral dissertation of Terry Winograd, about which I will say more later. We see here a computer program that has a small but noticeable fraction of the intelligence of humans.

MORRISON: What is that fraction?

MINSKY: I think the fraction is somewhere between 1 part in 10^6 and 1 part in 10. This program can fit into a memory of one million bits. I cannot conceive that it would take 10^{12} bits to hold a superintelligent being of this sort. I also cannot conceive that there could be one hundred times as much work to accomplish that as there has already been.

I won't guess how long it will take, because the time large technological projects take hangs on the enthusiasm of society. In one or two decades the possibility of artificial intelligence will become visible enough that society will have to decide, as it did in the case of space, whether to encourage it to advance. I am not sure what this says about interstellar communication, but I think Doctor Shklovsky's assertions are the kind I would make if I thought about it more.

GOLD: I would like to refer to the possibility of another phase in the technological revolution of our time which would change some of the discussion, not all of it, in a substantial way: Everyone thinks very much about the development of computing machinery, and no doubt that is a very important thing, but one ought to think of the development of just very large and powerful machines, mechanical machines of enormous size and power.

At the present time, we still deal with machinery on a scale which is determined by the human brain. We are not yet one full generation of machines removed from the brain. We make machinery like automobiles or road-making machines, or whatever, on a scale that is related still to the human frame through one agency or another. We have hardly removed ourselves from it. It has to be assembled by men or is at the present time assembled by men, or is made for purposes that are related to the size of the human frame, like the automobile.

But when a little time has elapsed, we will probably remove ourselves from that restraint and we will expand the scale of our machinery. The moment that we go in that direction I think we can immediately proceed in a huge way. Because the moment you can make a very much bigger machine, then you can also dig for ores and you can also make very much bigger machinery for production of metals and for every phase of the production of the machines themselves. There is very much positive feedback in such a situation, and the moment we take off in that direction we will go a long way.

From that, I would see that we would regard construction possibilities on scales that at the moment we don't even dream about. It would be done in the first place chiefly because such things may be possible for human habitation, such as taking water over really great distances—taking the Mississippi to California, if you like, or whatever it is, because you have digging machines of that scale. But it would also allow you to search for ores, for minerals, not in the first 2 or 3 kilometers of the earth's crust but in the first 60 kilometers of the earth's crust —you would be able to dig holes on that sort of scale.

In that case, the discussion we have had so far as resources are concerned is, of course, totally changed. I know the nature of the exponential function that Platt was referring to, and in those terms you change everything about every two years. Nevertheless, giant machines change it in quantity in a big way; suddenly you have 60 kilometers of ground to plow over if you want to. You change completely the question of pollution, for example, because you are developing new means of avoiding it.

You change completely the question of food supply because I have no doubt that in that era of big machinery, artificial foods of all kinds will become possible.

I believe huge change should be contemplated. In fact, I don't understand why we haven't moved in that direction yet. But when it comes, instead of the complications, we should think of the kind of capability that we would then acquire.

MINSKY: I would like to add to Doctor Gold's remark the possibility of very small machines, which I think will also begin to exist in the next decade. Our laboratory is working on them. Such machines entail the same kind of extension; for example, with respect to power we should be able to fabricate square kilometers composed of tiny solar cells, very delicate devices.

One of the main objections has been keeping them clean, but one could develop little insect-like machines to run over the surface and clean and repair, and I think Tom Gold's point is very well taken that the pollution and other crises that are mentioned here are best handled by an aggressive technological expansion.

THE NUMBER OF ADVANCED GALACTIC CIVILIZATIONS

$$N$$

SAGAN: I would like to draw some consequences from equation (1) and our deliberations to this point in the discussion. Some factors we felt we knew relatively well; others, very poorly. Let's put in the numbers we have been talking about and see what the implications are. We have explicitly mentioned a value of R_*, the mean rate of star formation over the lifetime of the galaxy, as about 10 per year.

The fraction of stars with planets, f_p, seems, from the presentation of Doctor Gold, to be of the order of unity, perhaps a half, a third, a quarter, something like that.

I tried to cover in my talk the number of planets per solar system where conditions are not so severe as to exclude life. I would guess it is somewhere around one. In our solar system it is certainly several. There are at least some hints that the intrinsically fainter main sequence stars have their planets closer in than we do in our solar system. And such stars are older than the sun, allowing more time for biological evolution on their planets.

What we have been debating most vigorously is the product of f_l, f_i, and f_c, the probability of the sequential emergence of life, intelligence, and advanced technology. This is where the question of subjective versus statistical probability arises most poignantly. Doctors Crick and Orgel felt that f_l might be near one, but that the uncertainties were so great that f_l might be much less. On the other hand they suggested that f_c is very likely to be ~ 1. Doctor Platt and I felt that f_l was near unity with somewhat greater reliability, but Doctor Platt and perhaps Doctor Lee were concerned that the evolution of language and the domestication of fire might be sufficiently unlikely to make f_c small. Doctors Minsky and Hubel have discussed the number of elementary neural units necessary for f_i to be near unity—it does not seem very large, and the selective advantage of intelligence is enormous. We have tried to distinguish clearly between the likelihood of the development of man, and the likelihood of the development of a very different being who is his intellectual equivalent or superior. We have also tried to stress those factors, requiring further study, where the time for their appearance may exceed the time available. The extreme range of all estimates for the product $f_l f_i f_c$ runs from unity to some extremely small fraction. But my sense of the average estimate of this subjective probability by the symposium participants is ~10^{-2}.

If we put in these numbers (and we are, of course, at liberty to put in any other numbers we want), we find that the number of extant civilizations in the Galaxy is $10^{-1} \times L$, with L measured in years. That immediately means that, even without estimating the lifetime of an average civilization, we have estimated the rate of formation of such civilizations in the Galaxy: one is formed about every ten years.

If we are talking about technological civilizations capable of interstellar contact, we are only a few times ten years old, so even under these assumptions, which some would perhaps say were optimistic, we are forced to conclude that there is no civilization in the Galaxy with which we can communicate which is as dumb as we. All other communicative civilizations are going to be substantially in advance of us. I think this is a point which has very serious implications for questions of contact, to which I return later (Appendix C). I stress it is independent of our estimate of the mean lifetime of Galactic civilizations.

Finally, N depends on what we choose for L. Remember L is some very crude mean of the lifetime of all civilizations in the Galaxy. Doctor Platt proposed, as others have, that there is only a small fraction, perhaps 1 percent, of civilizations which solve the problems which beset our age, but that those civilizations have very long lifetimes. With an extremely advanced technology, as Doctors Shklovsky, Dyson, Minsky, and others have stressed, problems which beset emerging societies such as ours will be solved. Such societies may be no longer limited to the planets of their origins. If such societies, stabilized against self-destruction or loss of interest, reach geological or stellar evolutionary time scales, that would imply that the appropriate value of L is $10^{-2} \times 10^9$ years or 10^7 years. The conclusion from our choice of numbers in equation (1) would then be that $N \sim 10^6$; there are a million technical civilizations in the Galaxy. This corresponds roughly to one out of every hundred thousand stars. Assuming these civilizations are distributed randomly, it follows that the distance to the nearest one is a few hundred light years. With any numbers like these the long-lived very advanced civilizations dominate the picture utterly.

But all our arguments about the value of $f_l f_i f_c$ pale before our uncertainties in L.

If we assume a pessimistic scenario, as reading the daily newspapers does not always discourage us from doing, we could assume that L on the average is decades. Then, you see, equation (1) implies that the number of technical civilizations in the

Galaxy is one—us. If civilizations destroy themselves shortly after arising, there may be no one for us to talk to but ourselves. MAROV: Doctor Minsky spoke of the inevitability of passing on to technological creatures and the inevitability of choice between the Golden Age of Doctor Stent and a transition to relatively few but highly organized technological creatures. Doctor Lee spoke of the evolution of intelligence on Earth, of the line of development which resulted in man as we know him today. We are thus speaking of a certain trend in the development of civilization, in the evolution of *Homo sapiens*. However, nothing was said here as to whether man today has reached his limits, the limits of his development, mental and physical, say from the standpoint of the possibilities of his brain to store far more information, to transform it in an optimal manner, in a limited time and so on.

It would be interesting to know whether anthropologists preclude at the end of this line of development of the intellect the possibility of a qualitatively new stage in man's development. We can only speculate how many of the previous branches in man's development were blind, that is, did not give a stable species— whether they were impasses. Possibly the development of a biologically intelligent creature such as man is today will also have a dead end and will give way to what Doctor Minsky spoke of, a new branch in the development of a technologically highly advanced civilization (even if mankind will avoid such catastrophes as the nuclear catastrophe, pollution, and so forth). But I wonder then whether the concept of the development of technologically intelligent creatures solves some problems facing biologically intelligent ones. Cannot even the computers of the fourth generation expected today be close to such a highly developed network of logical connections that impart to them the shortcomings inherent in man today—e.g., such contradictions as ambition, greed, envy? As an example I could here refer to a very conclusive situation in computer-astronaut interaction as drawn by A. C. Clarke in his *2001: A Space Odyssey*. The expected self-production of highly efficient machines and technological advances is scarcely a remedy to all our problems in CETI. Another possible trend in such self-production could be a degradation into a society of robots. To make this clear let us imagine we have a collection of robots who have decided on self-improvement. In that group there are attempts to solve certain mathematical and logical problems with emphasis on self-improvement by methods of self-replication, so that he who does this in the worst manner must be dismantled and reassembled in a more perfect manner. However, if we are dealing with a sufficiently highly organized (technologically intelligent) creature, evidently such creatures must develop a set of thinking abilities corresponding, say, to the interests of self-preservation. That could, evidently, give rise to a certain elite, richly conducive to the dismantling and reassembly of units at the same level or a lower level in order to preserve their own power. I would call this something like robot chauvinism. A situation similar to that has been used (and quite reasonably) in contemporary science fiction fantasies.

I would like to say on this score that our assessment of L as 10^7 years seems to be, while plausible, rather optimistic an estimate. I should nevertheless add that because both progressive and regressive tendencies appear to be intrinsic to intelligent society, the actual level of its development could undergo different phases. Then biological transformations or transformation of technologically intelligent creatures, would be represented as a process of civilization development in lines as pulsating from a decline to a flourishing state. In that case, in equation (1), if we apply it to estimate N in a limited area, a limited region of the universe where we are conducting our research, we must probably introduce a factor taking into account the pulsating character of the evolution of civilization.

LEE: I want to discuss several aspects of what might loosely be called SCETI, the Sociology of Communication with Extraterrestrial Intelligence, particularly the problem of instability, and the extremely complex issues of diffusion and the evolution of contacts. These issues may have a bearing on the values we assign in equation (1).

Let us consider L, f_c, and their effects on N, the number of technical civilizations. We are having great difficulty estimating the magnitude of L. The estimates vary from less than 10^2 to more than 10^9 years. We have the short L, the long L, and we have a cyclical L of a civilization that waxes and wanes, losing and regaining interest in communicating through time.

This difficulty should not surprise us because we have already said that intelligence is an unstable adaptation. The problem of instability is illustrated in the following facts: For 4×10^9 years in the development of life, things were OK; for 2×10^6 years of intelligent life things looked OK; but after only twenty-six years of technical atomic civilization we have already come very close to destruction.

How can we get a handle on this problem? One way is to think about various planetary environments and the rates of growth of life forms that are possible within them. Rates of growth could be measured in the number of new species to appear per million years of evolution. A rich atmosphere will allow for more trials or more species than a poor atmosphere or a poor environment.

This perspective raises the possibility of a planetary environment that makes for evolution that is *too fast* as well as too slow. In the search for life on other worlds, we have tended to think in terms of the minimal conditions, the lower limits, for the development of life. We should also think about the maximum conditions. There may be a number of environments that are so rich that species in them rapidly flourish and die out.

I am reminded of an eloquent phrase of Professor Shklovsky's where he speaks of a civilization that lives and dies in the space of a day like a butterfly. There is some indication that our planet may be such a case. Is it possible that the existence of fossil fuels and our exploitation of them during the Industrial Revolution helped us to reach the takeoff point too rapidly, long before the social organization had an opportunity to evolve the necessary checks on destructive forces? In this perspective, an instability can be redefined in terms of explosive growth and at least here we can visualize the conditions for life as being on a normal distribution curve with the optimum conditions being somewhere in the middle. Below the minimum no life occurs; above the maximum life expands too rapidly and collapses upon itself.

There is one additional large factor which may have an affect on equation (1), and that is diffusion. From our terrestrial experience, we see a tremendous impetus for diffusion, that is an outward expansion of powerful societies. Wherever we have civilization, we have imperialism.

Diffusion has several interesting properties for our search. First, diffusion from A to B places B in a subordinate position to A. The contacts are rarely symmetric, though I think this is a short-run phenomenon and is reversible.

Second, diffusion preempts invention. When Spanish civilization spread to Aztec civilization, that precluded the continued independent historical development of the latter people, who merged their destinies with the former. So we on earth will merge our destinies with our contacting ETI. Taken further this means that even our first encounter with ETI may put us in contact with more, perhaps many more, than one extraterrestrial civilization.

Third, diffusion has the effect of stimulating ETI, that is, increasing the number of technical civilizations. The reason for this is that a communicating technical civilization may influence others to become communicative. This has a powerful multiplier effect on f_c.

At this point, I think it is necessary for us to introduce or to clarify a distinction between primary-evolved civilizations and diffused civilizations. By "diffused" I mean a civilization that has been helped across the boundary of intelligence or across the boundary of technical civilizations as a result of direct stimulation from a "donor" extraterrestrial civilization. Of course we believe that we on Earth are an example of the first kind (primary), but we cannot exclude the possibility that we are an example of the second (diffused).

Therefore, in the future evolution of our contacts, we should think about the sociology of CETI and by this I mean the social structure of the communication between technical civilizations. We have emphasized the energy aspects of Type II and Type III civilizations. I would like to say something on the communicative or social aspects.

One possible scenario for the evolution of a technical civilization may go as follows: Stage 1 is the primary-evolved stage. In Stage 2, the civilization makes contact with ETI and becomes a a communicating civilization. The equipment built at this stage is largely if not solely for reception rather than transmission; later in Stage 2 equipment to transmit back to the donor ETI may be built. At this stage we are absorbing more than we are giving. In Stage 3, we become an exporting technical civilization by building broadcast equipment to reach others and we begin to give more than we receive. And Stage 4, we enter a highly advanced experimental stage of stimulating the development of ETI on other planetary systems by direct contact.

Stages 1 and 2 correspond roughly to Kardashev's Type I civilization, Stage 3 to his Type II and Stage 4 perhaps to his Type III. On earth we are presently at Stage 1 trying to get to Stage 2 but since both Stages 1 and 2 are silent we are seeking to contact a Stage 3 "exporting" ETI. Eventually we may find out what prior relation if any we have had with a Stage 4 civilization.

Finally one of the most difficult immediate problems we face in achieving contact with ETI is our limited terrestrial time-perspective. It is clear that communicating technical civilizations must be very long-lived, and this implies that the organisms, if we can think of them as such, are individually very long-lived, or that they are highly organized and have a strong sense of continuity, or both. The possibility has been raised that there are many technical civilizations based on short-lived organisms that don't have enough time or enough sense of the scale of the operation to get into contact.

On the other hand, there may be cultural evolution in this direction and the desire to make contacts may move us in a more time-binding direction, so that CETI may have the effect of lengthening L simply because if we receive a communication we want to stick around to find out what the reply is to our communication even though the time involved may be measured in centuries.

I think that one contribution the social science community can make, both here and abroad, is to develop programs of research into this question of time-binding and how we can develop institutions that have a longer life span than those currently conceived.

VON HOERNER: I want to discuss estimates about the *distance* between neighbors, some *crises* of development, and the *duration* of a search. The purpose of these estimates is not just to derive some uncertain statements about other civilizations; if we want to establish contact, we must have estimates for many quantities, estimates which are needed for guiding our actions.

For calculating the *distance,* we should know how frequently intelligent life occurs. But having no knowledge about any other civilizations, we must use basic assumptions instead; I will use the following two: "Nothing is unique," and "Nothing lasts forever."

"Nothing is unique" means we should assume that the only

case we know, life on Earth, is about average and nothing special. Can we do statistics with $n = 1$? Of course we can, but we should know the limitations. Having $n = 1$ yields an estimator for the first moment or average; there is nothing wrong with that. But it does not yield estimators for the higher moments, which means we do not know the mean error of our estimated average. Furthermore, in our special case there is a bias, because the whole estimate can only have been made where (some) intelligence exists, not on a barren planet or without one. In summary: assuming that we are average has the highest probability of being right; but we have no idea of how wrong it may be, regarding both the statistical and the systematic error. Actually, assumptions of uniqueness have always turned out wrong in our past: China was not the center of the earth, the earth not the center of the universe, our faith not the only one, and so on. Thus, the best we can do is to assume that we are average and leave a fairly wide margin for the error.

Some astronomical estimates show that probably about 2 percent of all stars have a planet fulfilling all known conditions needed to develop life similar to ours. If we are average, then on $\frac{1}{2}$ of these planets intelligence has developed earlier and farther than on Earth, while the other $\frac{1}{2}$ are barren or underdeveloped. We call $f = 0.01$ the fraction of all stars where life and intelligence are at least as far developed as on Earth, and $D_s = 1$ parsec $= 3$ light years the distance between neighboring stars. The distance between neighboring forms of higher life and intelligence then is

$$D_i = D_s f^{-1/3} = 14 \text{ light years};\tag{4}$$

or, with an uncertainty of, say, $f = 10^{-1}$ to 10^{-3}:

$$D_i = 6 \text{ to } 30 \text{ light years.}\tag{5}$$

"Nothing lasts forever" means we should never assume that our present state of mind, with its strong dominance of science and technology, is the only and final goal of all evolution. It will be only one link in a long chain, to be surpassed by (unpredictable) other interests and activities. Technology, then, will probably be maintained just to keep things going, but without attracting any individual genius or big public money. This termination of the technical state I would call the "change of interest." In any case, we should assume a finite longevity L of the technical state of mind; but there might be other terminations, too, like the following ones.

Technology has its dangers, resulting in several *crises,* which might result in a termination of the technical state or of the whole race. First, we have just began to feel the impact of the "population explosion," with overcrowding, pollution, and diminishing resources; the most crucial part of it is the breakdown of social instincts and patterns as demonstrated in animal experiments. This crisis must be quite general, since every member of a successful species must have a strong inherited urge to fight against his own death and to raise many healthy children, which makes the use of medicine emotionally much easier for diminishing the death rate than the birth rate.

This crisis is more severe than usually realized, because (a) our population increase is not exponential as mostly quoted. The growth rate actually is not constant but proportional to N, and the best-fitting curve for the past 2000 years predicts an infinite population in 2026, after only 54 years. (b) Even with perfect technology, the problem cannot be solved by interstellar expansion. Frank Drake once suggested over coffee break that the finite speed of light sets a limit, and this turned out to be right: if we populate all habitable planets within a sphere of increasing radius, such that the volume of this sphere increases with 2 percent per year (our present growth rate), then the limit is reached when the radius of the sphere increases with the speed of light.

The resulting numbers are amazingly small: the limiting radius is only 50 parsec = 150 light years; within this sphere are 30,000 habitable planets, but starting with 1 today, and increasing with 2 percent per year, it takes only 500 years to populate all of them as densely as our earth is now. After that, a growth rate of 2 percent per year cannot be maintained without overcrowding, and we are back to the same problem. Or, consider the circum-stellar spheres as once suggested by Freeman Dyson, where a highly advanced civilization takes some of their larger (and otherwise useless) planets apart and builds with this matter a sphere around their sun, in about their own orbital distance from it. This allows them to make use of the total energy output of a whole star (Kardashev's Type II Civilization), and it increases their Lebensraum by 10^8, a tremendous factor. But again, with 2 percent per year a factor of 10^8 is used up after only 1000 years. Thus, even a perfect technology with the ultimate method cannot solve the problem; it can only delay it by 1500 years.

(c) Birth control cannot be left voluntarily to the reason of the individual. As mentioned by Garrett Hardin, this would lead to a genetic self-elimination of reason.

Second, another crisis we just have entered is "self-destruction." Exactly 10 percent of all our human effort (gross national products) goes into fabrication and development of weapons; and our accumulated destructive power at present amounts to 10 tons of TNT per person, resembling a round ball of dynamite 2 meters diameter for every living person—grandmothers, babies, and all. Furthermore, the present stalemate of the big powers cannot remain stable forever, as pointed out by von Weizsäcker: if weapon systems get obsolete every seven years, say, to be replaced by improved ones, then there is at least a small chance every seven years for a war; these chances accumulate, of course, and with a single chance of 10 percent, for example, war becomes more probable than peace after only 45.6

years. This crisis, again, must occur quite in general, because every member of the species dominating a planet must have a strong inherited urge to dominate, to fight all competitors by all means.

Third, a further crisis to be predicted is "genetic degeneration." It began when medicine began to eliminate natural selection while mutations still went on. It will need a longer time, some thousand years, to become critical. The only way out is artificial or at least guided breeding (needed for reasonable birth control anyway) and some day we must start with it. This is more critical than it may seem at first glance. For example, who decides, and how, which human qualities are to be favored, which part of the population is to be discontinued? Just try to imagine all the fights, the lobbies, the intrigues, and frustrations. And irreversible mistakes are possible.

Fourth, to overcome all these (and probably many more) crises, all surviving civilizations must have developed strong means of regimentation and stabilization, which may lead to stagnation in many cases, the crisis to end all crises. Finally, with guided breeding, "irreversible stagnation" becomes possible.

In summary, the finite longevity L of the technical state can be defined by a change of interest or by one of several crises. If we call $T = 10^{10}$ years the age of our galaxy and its oldest stars, then the distance between neighboring technical civilizations is approximately

$$D_t = D_i(T/L)^{1/3} = D_s(T/fL)^{1/3}. \tag{6}$$

But for estimating L we do not even have $n = 1$; we have just started the technical state and do not know how long it will last. Instead of an estimate, we can only make a free guess, for example, $L = 10^4$ to 10^6 years. With $f = 10^{-3}$ to 10^{-1}, the total range for D_t then is

$$D_t = 140 \text{ to } 3000 \text{ light years.} \tag{7}$$

My own guess would be $f = 0.01$ and $L = 10^5$, which yields about $D_t = 200$ parsecs $= 600$ light years. This, finally, is the distance to be bridged by the signals of interstellar communication. The waiting time for answers, then, is 1200 years, which shows two things: (a) this is a communication between whole civilizations, not individuals. (b) It is either a one-way communication, as was the one from the ancient Greeks to us; or it is question-and-answer, but only so if the time scale of development is more than 1000 years, much slower than our own rather hectic one.

What is the *duration* of a search, needed for detecting the first extraterrestrial signals? There are three different cases. In case A, we are among the first ones who ever try, and so are our prospective partners. We are all on about the same level, should all contribute the same effort, and we should transmit and receive. In case B interstellar contact has already been established long ago. Communication mostly tends to create a common culture, and we may expect a similarity to the origin of life itself: it may be difficult and time-consuming to get it, but once originated, it then has a tendency to continue in time and to spread out in space. Thus, we may expect a galactic community, much further advanced than we are, trying to attract the attention of future new members, probably by what I called contacting signals. In this case we should just receive, leaving the burden of a powerful transmission to the more advanced partners. We will have success once we have guessed the right method and built enough equipment. The duration cannot be estimated in advance; it depends on how long it takes us to get enough clever and dedicated searchers. Finally, in case C, there is plenty of advanced technology, but nobody is interested in talking with us. In this case we can just search for leakage from their local broadcast; we know what to look for but expect only very weak signals. After estimating distance and power, the duration for achieving a given signal-to-noise ratio then is proportional to the inverse square of our receiving area. The duration thus can be shortened by paying a higher price.

A fairly good estimate can be given for the minimum duration in case A. Suppose that a fraction F of all stars develops higher life that goes once through a communicative phase, transmitting and receiving, of duration τ. Then the distance between simultaneously trying ones is

$$D_a = D_s(T/F\tau)^{1/3} \tag{8}$$

and for success (getting an answer) one must continue for at least a duration of

$$\tau \geq 2D_a/c, \tag{9}$$

where c = velocity of light. Both equations together then give

$$\tau \geq (2D_s/c)^{3/4}(T/F)^{1/4}. \tag{10}$$

Fortunately, there is only one uncertain quantity, F, entering only with the power $\frac{1}{4}$. If $F = 0.001$ to 0.1, say, then

$$\tau = 2000 \text{ to } 6000 \text{ years}, \tag{11}$$

and the distance to be covered is

$$D_a = 1000 \text{ to } 3000 \text{ light years}. \tag{12}$$

These are minimum values, and actually it might take a lot longer. But even with adequate technical means, and with an ideal method, it takes at least 2000 years. The nice thing is that anybody else, on any other planet, can make the same estimate and will get the same answer. On the other hand, I think that case A is less likely for us because our sun is not among the oldest stars, which are about twice as old. If we really are average, the first communications will have been discussed already 5 billion years ago.

Finally, I would like to add that interstellar communication may well be of crucial importance for the development of civilizations, just as speech is to the development of individuals. Furthermore, it gives all the advantage of competition (avoiding stagnation) without the worst of its dangers (mutual killing).

KARDASHEV: Many of the speakers have honestly confessed that they have no certain answers to these questions and will not have them in the immediate future. Such a situation arises very often in many sciences and it is most useful in such cases to adopt several working models which can then be elaborated upon or rejected. Evidently in this case, too, it would be most useful to adopt some formalized model of a civilization which we, after discussion, would adopt for the near future, and have several such models so that in the future we could see what our models lead to.

From what I understand here, I feel that one of the most important problems in the development of living organisms and civilizations is the process of the reception, processing, and analysis of information. I submit this to you for criticism. I think that this process of natural selection in the evolution of civilization must be linked with the analysis and processing of information. If this is so, then proceeding from information theory, we can endeavor to produce several models that could be extrapolated into the future so that we could in this way see what happens in different situations, adverse and favorable, what we can expect.

Therefore, I think we can conclude that civilization, man and perhaps higher organisms, are systems that receive and process information. Thanks to natural selection, they develop the principle whereby they attempt to process and utilize a maximum amount of information. Can we offer this as a basis for describing many phenomena, specifically for describing phenomena involving distant civilizations?

Our endeavor involves extrapolation in some way into the future. We must reach some understanding here, since we do not know what laws of nature may operate, laws which are unknown to us, and which may link the critical civilization that we are trying to reach. Everyone will agree that, in addition to the information I mentioned, such inevitable attributes as an increasing scale, energy, space, and time will characterize advanced civilizations. That will be discussed in greater detail later. But I would merely like to say now that evidently the most important factor here is the process of the utilization of information in the initial period of development of a species, for its self-improvement. Perhaps at a higher stage, perhaps in the future, the set of data could affect the development of science, the development of art, the development of technology. Can this be extrapolated? Are we justified in doing so?

The second question I would like to touch upon is the question of equation (1) which initiated our discussion. This formula, too, is a useful model that can be elaborated upon. I agree that it must include a feedback effect. This feedback effect may be negative or positive, and that would have a substantial influence on our estimates.

The feedback effect has the result that contacting a civilization could influence the number of civilizations. There could be a sharp increase, a rapid development of those contacted civilizations. I think the optimum figure here, the most optimal version, would be the merger of two civilizations into one, and that ultimately this could reduce the number of civilizations in each galaxy to one—but one very extensive in its attributes.

Modern astronomical data imply that the universe corresponds to an open Friedman model. If so, the universe has an infinite number of stars and galaxies. Consequently equation (1) has infinite factors, and the total number of civilizations in the universe is infinite. But for communication problems, the number of civilizations from which signals may reach us is important. This value is finite, but increases with time to infinity (proportional to T^2, where T is the age of the universe). For example, the universe model with density $\rho = 2 \times 10^{-31}$ grams per cubic centimeter (this value given only by the galaxies is a minimum density) gives 10^{15} galaxies inside our horizon instead of 10^{11} as mentioned at the beginning of the discussion.

My last remark concerns the problem of the immediate future, which was also discussed here. It is my opinion, as is the opinion of several others, that there is a great future in computers and the consequent increase in data processing. This is in good agreement with the principle of the maximum increase of data which I tried to formulate here. With respect to using the maximum resources, we must take into account the expected emergence of man into space which may greatly enlarge our potential and in some way help us to avoid the difficulties which have been mentioned here.

IDLIS: I would like to underscore the fact that individual intellect, natural or artificial, constituted of some elementary units, must have finite dimensions and relatively small dimensions. This means that such an individual intellect has a finite intellectual capacity, too. It does not mean that we cannot assemble a system of higher intellect, but it means that such a system will have to have a certain hierarchical structure. We observe in human society that there are creative groups, official and unofficial, visible and invisible, and these can also be observed in the computer field. These hierarchical structures are considered in sufficiently general terms in my treatise *Mathematical Theory of the Brain and Optimal Size*. This is a book on the mathematical theory of the optimal organization of work, and in this book I consider the optimal structure of working groups in science. But the concepts are, I believe, applicable not only to a case where we are dealing with live individual research workers; I feel that the same may apply to an hierarchical structure, including a system consisting of computers.

I would also like to mention the problem of defining a cognitively developing civilization. In order to study anything, we must, of course, either know the subject or be able to define it. In the case of CETI, we know nothing; we are forced, therefore, to make definitions. Engels said definitions do not of themselves mean anything in science. Nevertheless, they are an essential premise.

As a working definition of a cognitively developing civilization, I suggest we use that offered by Doctor Kardashev: a highly stable state of matter capable of assembling, analyzing, and utilizing information for the maximum knowledge about the environment and about itself, and conducting maintenance reactions.

Relative stability is a property of all real objects. It is a necessary condition for their existence. A passive reflection of current conditions of existence corresponds to inorganic matter. An active reflection, with due consideration for the real past, is a property of living nature and an advanced reflection—that is, a knowledge of the future—is a property of intelligence.

In this respect, the intelligence, the civilization characterized by intelligence, differs from all previous forms of matter. The intellect has unlimited possibilities of development in this context. What do we mean by saying that this or that civilization is making progress? The concept of a progressively developing civilization boils down to the development of that civilization in science, to the consecutive solution of topical scientific problems. The number of current problems requiring solution must grow exponentially. If there is a need to solve only one problem, that simply means that we have described as a solution what is merely a stage in the examination of a problem. This is only a quasi-solution. If the solution of one problem does not lead to the need to solve new problems, then we speak of a quasi-problem.

This is not merely a philosophical postulate; it is a mathematically proved theorem, Gödel's theorem, which implies that no field of science can be represented as an inexhaustible list of accidents, determining certain concepts. There will always be, as Gödel demonstrated, a situation formulated in terms of these concepts which cannot be proved or refuted on the basis of our setting up axioms, and sooner or later it will be necessary to regard that postulate or its negation and then together in a dialectical synthesis it will be necessary to refine them in association with new axioms. It is in this way that the axioms of science expand without limit.

In order that a civilization could be called progressively developing, it must deal with all its topical, current problems. It must solve them in the proper time, not putting them off, for as a civilization selects some problems as being more substantial than others, it will later be solving a smaller and smaller share of the real problems facing it. Civilizations will eventually be solving one problem only, that of their existence.

To sum up, then, for a civilization to develop progress it is essential that that development be exponential. Ever since science first arose, ever since, I would say, Newton's time, the development of science has indeed been exponential, with a doubling every ten or twelve years.

So that development should be exponential, we must overcome a number of obstacles. Von Hoerner says that cosmic expansion into space cannot make exponential development possible for more than another few thousand years. What happens then? Is there a possibility to follow exponential development further? We must extend civilization not only in our universe but across through all the elementary particles into neighboring coexisting quasi-closed systems of worlds; in recent years a number of authors have developed the hypothesis that particles which we regard as elementary may indeed contain hidden macrosystems. Such an extension into them may one day take place.

DYSON: The references that were made to Gödel's theorem I believe were mistaken. This is in connection with the logical structure of the message. In fact, Gödel's construction provides precisely a message by which a language is able to make statements about itself. There is no logical impossibility in this.

PODOLNY: Historians find themselves in a more difficult position even than astronomers when discussing CETI. Astronomers *have* Barnard's star. We don't even have that. For a long time we hoped there would be a civilization on Mars to look into, but that possibility seems to be vanishing quickly and that leaves us only with the earth.

The question as to whether there need be a technical society at all seems to have been answered satisfactorily, but we see that even with Homo sapiens who were capable of creating a technological society, even here there is a large group of people who advanced toward such a society so slowly that we cannot even say this would continue as progress toward that goal. Take the Australian aborigines—it is very hard to say in what direction they have been developing in their more recent phases of development.

However, there is evidence in support of the argument that a technological society is attainable by any earth-type civilization. This is supported by the argument that Japan, cut off from European civilization for a long time, nevertheless developed, with a relatively small time lag, quite rapidly and in a very similar direction, and this leads many Japanese scholars in the Soviet Union to regard this as an explanation of the spurt by which Japan overtook the advanced European civilizations and advanced to one of the leading civilizations in the world today.

Some historians also note the psychological effect and would, for centuries, speak of "them" and "us." Now, despite all the events of the past century, humanity is more and more feeling itself to constitute one whole, and we are all coming to be "we." To seek a "they" on some other planet requires some courage, and that, I think, accounts for the interest of a large number of people in CETI.

We must look at ourselves through the eyes of an extraterrestrial neighbor who can observe us from the side without being weighed down by our history and who can see with such a gaze what it is in our nature that is determined by biology or by our long and tortuous history, and what is inherent in our intelligence and our senses which we value so dearly.

There is another part of mankind that is also interested in CETI but the problem there acquires a different aspect. Most of those people seem to have lost faith in the possibility of any supernatural element intervening in their life, and they seek some substitute for that supernatural element. I think this accounts for a considerable share of the interest in the matter that has brought us all here together. I recall, in particular, one American book about unidentified flying objects that begins with the categorical statement that our civilization should come into contact and begin negotiations and ends with the statement that our governments are concealing the fact that they *have* entered into negotiations with other civilizations. That is a case in point of what I was saying.

To examine its implications we have set up a commission. We gave it a somewhat fantastic name, the Committee for Contact with Extraterrestrial Intelligence. Its composition differs considerably from the composition of our gathering here. Here we have astronomers and physicists prevailing; there we have anthropologists, historians, and psychologists as the prevailing contingent.

The first and immediate task of that committee is the examination of certain problems for purposes of enlightenment, I would say; to examine the arguments of those who say that contacts have already taken place somewhere in the past and that there are traces of visitations on our planet by representatives from outer space.

We have published several articles attacking these sensational contentions. Nevertheless, at the same time we have found a certain number of cases that do merit certain study and I will dwell upon one of these cases. This is a case from a document published in 1842—that is 130 years ago when there was no discussion about CETI. That, of course, makes it all the more reliable. This is a report of a monastery clerk in Northern Russia addressed to a high dignitary of the Russian Church who reports that on August 15, 1663, there was a visitation of the earth between 10 and 12 hours from the clear skies. A sphere appeared, about 40 meters in diameter; from the lower part two rays extended earthward and smoke poured from the sides of the vehicle. The body disappeared and reappeared again, again disappeared and reappeared, changing in brightness in the course of these peregrinations. The phenomenon occurred over a lake and lasted for an hour and a half. At the place where the sphere touched the water, a brown film appeared, resembling rust. The phenomenon was observed by two groups of people. Some watched it from the church; others from a boat which happened to be in the middle of the lake.

About ten years ago an Australian radio astronomer, R. Bracewell of Stanford University, suggested that there may have been a visitation of the planet earth by automatic probes in the past. One possible explanation for the phenomenon that I have referred to is that one such interplanetary probe did indeed appear on our planet. Naturally, I do not insist on this explanation and there may be many natural effects, many natural phenomena which simply have not occurred since.

I will not go into other examples that we have before us, but our committee has examined the methods by which it might be wise to scrutinize our folklore in search of evidence of such visitations. The method has not yielded any results to date but the approach proposed by our psychologists does deserve consideration, I believe.

I would refer to this historical fact: three thousand years ago or so, possibly slightly less, the Phoenicians traveled around Africa. The news of this would probably be regarded as a myth or would probably be questioned if documents that have survived did not state that in the course of their travels they observed the sky moving from the wrong direction, as they put it. We have every reason to think that this information was probably that which aroused the most skepticism among their contemporaries, but to us this is weighty evidence that such a journey did, indeed, take place. Because it is hard to believe that anyone could have invented such a description of the sky.

So, the human imagination, you see, is so constructed that people build up certain details as Leonardo da Vinci did in drawing a flying dragon. Man uses his imagination very broadly, but all the products of the human imagination are based on realities. It is our duty to find details that could not have been invented but which refer to realities.

ASTROENGINEERING ACTIVITY:
THE POSSIBILITY OF ETI IN PRESENT ASTROPHYSICAL PHENOMENA

DYSON: I have six points which I shall go through quite briefly. The first is: to hell with philosophy. I came here to learn about observations and instruments and I hope we shall soon begin to discuss these concrete questions. I am not competent to speak about instruments and, therefore, I will be very brief.

Point 2. Our primary concern is to observe extraterrestrial civilization if it exists, so we are biased in our search toward big technology, since big technology is, by definition, the sort that we can observe.

Point 3. If a society is very highly developed technologically, it must emit intense infrared radiation, not necessarily a planetary spectrum, but necessarily a large intensity of infrared radiation, whether or not this society wishes to communicate. Consequently, we should use infrared emission as a signpost indicating priority areas toward which we should direct searches by radio and other techniques.

Point 4. Any observing program should be primarily an investigation of natural objects. It should not be separate from the mainstream of radio astronomy. By looking carefully at the most interesting natural objects, we are guaranteed the reward of a harvest of important scientific discoveries. The chance of discovering artificial objects, if they exist, is an additional bonus. I fully support Shklovsky's dictum that every object must be assumed natural until proven unnatural.

Point 5. Much too much attention in our discussions has been given to planets. Planets may be a good place for life to begin, but they are not a likely place for the home of a big technological society. Instead, we should perhaps think more about comets. In our own solar system, there are probably between 10^9 and 10^{10} comets. That number is, of course, not well known but it is something of that order of magnitude. If you take 10^9, then the solar comets have the following properties: Their total mass is only 0.1 the mass of the earth. But their total surface area, which is, from the point of view of biology, much more interesting, is 1000 times the surface area of the earth. If you look at the mass of biologically useful material, that is, the carbon, nitrogen, oxygen, and hydrogen, as they are in the earth's biosphere, the available biosphere on the comets is 1 million times that on the earth. So I conclude that comets are likely to be the main home for life in the solar system. I mean not now but in the future.*

Point 6. There is quite a crucial question whether comets are a local phenomenon in our solar system or whether they pervade the whole Galaxy. If they pervade the Galaxy, then our conventional picture of interstellar travel is quite wrong. The distance between habitable oases in our Galaxy would be of the order of one light day instead of many light years. This would, of course, change the whole picture of how interstellar travel would look both to us and to our competitors. I conclude we should look for

* For a more detailed discussion of this and related proposals see Appendix D.

artificial signals among diffuse sources of radio and infrared emission rather than from individual stars, or perhaps in addition to individual stars.

GOLD: Do you believe that we should look at our present comets in the solar system in this context?

DYSON: That is an interesting proposal which I have never considered.

OLIVER: Why do you suggest that civilizations must of necessity produce large amounts of infrared radiation? It seems to me that the infrared radiation that would be produced by even a very much farther advanced civilization than ours would be negligible compared to the primary star. For example, in California, which has a very high usage of electricity, the power generation at the present time is only 0.1 percent of the sunlight falling on the state.

DYSON: What I am saying is that the civilizations which are observable to us will have this character.

OLIVER: But you are suggesting, are you not, that the infrared emission will be an observable characteristic? I am suggesting it is far down in the stellar noise.

DYSON: No, I am saying that the generation of large amounts of infrared radiation is not necessarily an accompaniment of a high civilization at all. Only if it occurs is it something that we can see.

MINSKY: Since radiation at any temperature above 3°K is wasteful and a squandering of natural resources, the higher the civilization, the lower the infrared radiation. We should look for extended sources of 4°K radiation. There should be very few natural such sources.

DYSON: I don't quite go along with this, but to some extent you are right.

GOLD: Is energy or surface area the bottleneck for emission? If the surface area for emission is expensive, then 4°K isn't very good.

MINSKY: Right.

BURKE: I would like to question the energetic argument that favors using the infrared, simply because of the analogy with recent progress in discovering remarkable objects in our own Galaxy. The pulsars emit all their energy, to first approximation, in the radio part of the spectrum. I think that Point 1 should be examined closely both as a warning and as a guide.

GOLD: Did you want to respond?

DYSON: That wasn't a question. It was a statement.

SAGAN: Harwit has discussed the infrared detectivity of astro-engineering activity (in Appendix E).

GINZBURG: Why do you see such advantages in the comets? If the comets we know are what you refer to, they seem quite unsuitable for civilization. If you want to change the comets completely, they will no longer be comets. Is it comets we know about or quite different artificial comets?

DYSON: I mean the long period comets that come into the solar system from far away and then disappear.

GOLD: Because of the material that they would provide?

DYSON: Yes. We know that these arrive in the solar system at the rate of one every two or three years and, assuming this has been going on for the whole history of the solar system, it means there must be a very large number of them loosely attached to the sun.

SAGAN: There is a number that may be of interest. One can ask, if each of the 10^{11} stars in the Galaxy had a comet cloud of approximately our number and dimensions, and if stellar perturbations were causing the loss of comets into interstellar space around all these stars at the present rate within our solar system, what is the mean time interval between the arrival of an interstellar comet on a hyperbolic trajectory in any of these solar systems? The answer is about a thousand years.

DYSON: Right. So the fact that we have never seen an interstellar comet is no argument against them.

TOWNES: I can understand your noticing that comets provide good raw material for civilizations, but with the particular sizes which individual comets have and the particular visible characteristics, I don't understand why they in their present form would be suitable for advanced civilizations. Could you explain this a little more?

DYSON: They have, of course, the advantage that you can get away from your government. But aside from that they have the advantage of just being abundant; they are the largest available living space that we know about.

GOLD: With the added problem that if they accommodate themselves for a long time to interstellar conditions, and then are perturbed to the inner solar system, then suddenly in a period of a year or so they would be imprisoned. Close to the sun you would have a huge ball-up.

DYSON: You just don't go close to the sun.

KARDASHEV: In dealing with extraterrestrial intelligence, we must concern ourselves with certain definite models; if we are considering a model of a supercivilization, that is, a civilization that is far ahead of ours, in looking for it we must take into account things we know nothing about. Many people think that nowadays in astrophysics we know a great deal about all objects. In my opinion, this is not so at all.

I would like to discuss one example of a phenomenon which might give us a clue as to how and where we should seek extraterrestrial civilizations if they take advantage of this phenomenon. What I would like to speak about briefly is the collapse of supermassive objects; my discussion rests on considerations that have been developed recently by Professor Novikov and by Academician Sakharov.*

Let us examine the position of an observer accompanying a large and contracting mass, larger than 1.5 solar masses. The state of such a large mass has now been studied very thoroughly. Contraction proceeds until the gravitational (or Schwarzschild) radius r_g is reached; $r_g \simeq 3(M/M_\odot)$ kilometers, where M is the mass of the massive object, and $M_\odot$ is the mass of the sun.

For an observer situated outside the system, the approach of a spacecraft to the mass will take on a definite appearance. As the spacecraft draws closer to the gravitational radius, all processes observed will be extended indefinitely in time; whereas for someone on board the rocket, the processes will take place in the usual time scale. This is a well-known effect.

Let us now proceed to examine what happens once the gravitational radius is reached. To begin with, the external observer will not observe this at all. The occupants (I hesitate for another term) of the spacecraft will reach the center in a period of time $t = r_g/c = 10^{-5}(M/M_\odot)$ seconds. After that, everyone will probably assume them to be lost, but some new models suggest they may remain very much alive. This is based on a model of the contraction of a large mass with electrical charges (Figure 11). Again in this case, the mass and the spacecraft contract within the gravitational radius, but contraction does not continue to infinite density. It stops somewhere not too far from r_g, and there is followed by an expansion. The maximum density must be at the moment of stopping. If the charge $\epsilon = \sqrt{6M}$, $\rho_{max} \approx 10^{18}(M_\odot/M)^2$, and $M \geq 10^9 M_\odot$, we have normal conditions, $\rho \leq 1$ gram per cubic centimeter. The massive object with the spacecraft expands once again beyond the gravitational radius.

* Soviet *Journal of Experimental and Theoretical Physics* (JETP), 59 (1970): N7, 262.

Figure 11. An observer in the vicinity of a black hole of radius r, mass M, and electrical charge ϵ. Consider the flight of a spacecraft into such a charged Schwarzschild sphere. If the charged body is contracting and the spacecraft is on its surface, then as $r \to r_1$ (where

$$r_1 = \tfrac{1}{2} r_g [1 + (1 - \epsilon^2/GM^2)^{1/2}]$$

and

$$r_g = 2GM/c^2)$$

an infinitely long time passes as observed by the outside universe. On the spacecraft all history of the future of our universe may be seen if $r \to r_2$, where

$$r_2 = \tfrac{1}{2} r_g [1 - (1 - \epsilon^2/GM^2)^{1/2}].$$

In the interval between $r = r_2$ and $r = r'_2$ the ship emerges from the Schwarzschild sphere and may examine this new space-time continuum. Near r'_1 it is possible to see all of the past of this new universe. Here, r_1 does not equal r'_1 and $r'_2 = r_2$, since near the end of the journey arbitrary conditions are in force, probably determined by the physical parameters in this new space-time continuum.

They return. But the main question is: Where do they return to? To the external observer, they will never emerge. In an infinite period of time they will still not have emerged. From this, we draw the conclusion that our space is much more complicated than it seems. Sakharov assumes that there is an infinite multitude of spaces separated one from another by infinitely large times (Figure 12). This provides us with a time machine which enables us to cover infinitely large distances in finite times, and which enables us to cover infinitely large time intervals in small proper times. This is rather abstract.

MORRISON: Yes, very abstract. It is a great trip for electrons but I would hate to send a spaceship on it.

KARDASHEV: Do not be afraid. The flight conditions may be normal. The density, radiation, and gravity gradient may be safe for life if ϵ and M are sufficiently large. Suppose we create a charged sphere which enables you to pass into another space in the in-

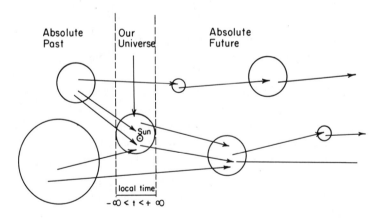

Figure 12. A representation of the isolation of our universe from other "universes" in what from our point of view is the absolute past or absolute future. The arrows represent possible traverses by spacecraft through black and white holes from one space-time frame to another.

finite future; for the spacecraft this will be measured in a proper time of microseconds. You then explore this new universe in your spacecraft which contains hotel, library, and laboratory. Then you repeat the process of collapse and pass on to the next object. We can also assume a similar transition from the past to the present. According to current terminology, objects coming from the past are termed white holes, and conversely objects through which we may enter the future are called black holes. It is important to point out some interesting features of such a journey. An observer may see all the past of the universe during a short time expansion of the white hole and he may see all the future while immersed inside the black hole. Figure 13 shows the redshifts for external and comoving observers. These considerations lead us to a model of Type III civilizations (Figure 14) [which are defined as commanding energy resources $\sim 10^{44}$ ergs per second]. I will, therefore, sum up by suggesting that astrophysicists and CETI investigators concentrate on white holes and black holes and on the processes that take place in their vicinities.

MORRISON: One question. What is the maximum tidal stress between one end of the hotel bed and the other at some point during this voyage?

KARDASHEV: The tidal stress depends upon the dimensions of the body relative to the gravitational radius. Such estimates have been made and everything seems to be quite satisfactory.

GOLD: I thought if the radius was only 2 kilometers or so, the stresses would be very large and a human being would certainly be drawn into a long, thin thread.

KARDASHEV: Of course, but for a very short time, you must remember.

GOLD: I don't wish to be a long thread even for a very brief moment.

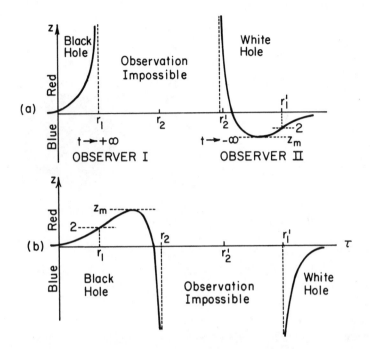

Figure 13. Schematic representation of black holes, white holes and mutual observability after Novikov and Sakharov. The frequency red shifts and blue shifts observed by external and comoving observers are measured by z; the maximum and minimum frequency shifts are shown by z_m. It may be possible to plunge into a black hole and reemerge through a white hole some infinite distance and time away, through which observation is impossible. Shown is the red shift for a journey inside a charged Schwarzschild sphere: (a) for an external observer tracking the spacecraft and (b) for an observer on the spacecraft examining his surroundings. Here

$$z_m \simeq \left[\frac{2}{1 - (1 - \epsilon^2/GM)^{1/2}} \right].$$

Compare with the caption of Figure 11.

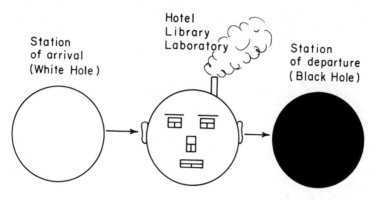

Figure 14. Representation of a Type III civilization according to Kardashev. The illustration is extremely schematic only.

DYSON: If you want room temperature and ordinary pressures, your supermassive object must be much larger than one solar mass.

KARDASHEV: Yes, it will be a quasar.

DYSON: Then the times are of the order of a day.

SAGAN: In case the tidal stresses are too large for the scheme to be manageable such black holes might nevertheless be relevant to our problem because I can imagine that they would provide significant navigational hazards; interstellar civilizations would certainly want to surround them with buoys to warn of their presence. The buoys may be detectable by us.

MINSKY: If the time of emergence is infinite, then the probability that he will emerge in any particular time is strictly zero, so no such event will ever be observed.

KARDASHEV: I think that the nearest white hole/black hole may be in the center of the galaxy; we almost know about this. Infrared observations suggest that near the center of our galaxy there is a very large cloud with a temperature of 270°K.

SAGAN: Lederberg (Appendix F) suggests that the center of the galaxy is the one obvious site for a Type II$^+$ civilization transmitter.

I still have the impression that even if no one knows what is happening in the vicinity of a black hole, clearly there are processes of great importance occurring there. There are large energy sources available and there are conceivably apertures to other places or other times. If that is correct, then there may be large technological apparatus deployed in the vicinity of black holes. If we had reasonable candidates for nearby black holes, perhaps Cyg X-1, then it might be wise to include such places on our list of candidate technical civilizations to examine by radio and other means.

OZERNOY: I would like to express one sobering thought about the possibility of the existence of technological systems near black holes, of which Sagan spoke. Through the radius of a quasar nucleus there is a flux of hard radiation in the form of ultraviolet and x rays of about 10^{12} rad per second. Any crystalline cell of a modern computing device is put out of action by irradiation of the order of 10^{14} rad. Even allowing for cybernetic devices with organic molecules in the polymer chains, such a computer will be put out of action in 10^4 seconds or less. We therefore have to overcome exceptionally adverse circumstances even to approach a quasar, let alone inhabit it.

SAGAN: Compared to the tidal stresses, the radiation field, especially for an object of stellar mass, is the least of our problems.

GOLD: Dyson once suggested that there was a kind of pseudochemistry possible in the outer layers of a neutron star with enormous density and with very short time constants. Frank Drake and I have wondered whether we could have a life form that took advantage of this kind of chemistry. I wonder whether Freeman would like to comment on that.

DYSON: No comment.

GINZBURG: The question I have been asked to discuss is this:

Can it be that on some remote planets (or comets, I must now add), as potential centers of ETI, there exist some different laws of physics? Obviously, if this question could be answered in the affirmative, this would have tremendous implications for all our estimates and would greatly change the nature of our discussion. But I use the words "I have been asked to discuss" for a very definite reason, not to saddle someone else with the responsibility for the absence of anything interesting in my communication, but because I had the feeling when asked this question that conventional physics is quite enough for us to deal with. It is rich enough and offers plenty of possibilities. I appreciate what Dyson said: Let's get down to business—get down to the matter at hand and examine the matter of the possibilities of communication, and so on. But I will do something different because that is my subject.

I have not found anything of exceptional interest but what I have found is food for thought. First of all, let's consider the perspective of the man in the street. Why must he believe that on a different planet the same laws of physics must operate? After all, we haven't been there. The simplest answer would be to say that we haven't been on Mars, either; possibly we haven't been on some small island in the Pacific; but we have no reason for assuming that the laws of physics there are any different, or we could make this assumption about different laws of physics with just as much relevance to these islands. The whole of natural science, actually, rests on the postulate that in the same conditions you have the same situations.

But what I think is not so trivial is how far we must extrapolate to generalize such conclusions. For example, in our galaxy we have about 10^{70} electrons and hydrogen atoms. Although we have investigated a quantity quite incommensurate with this number, we are quite confident that hydrogen atoms and electrons are the same everywhere. But there is a problem here.

Why are we so convinced that immaculate conception is impossible? We know that parthenogenesis is possible even in so complex an organism as a turkey; and I read in a magazine that some lady did claim to give birth to a child without a father—and she knew some genetics, that this could only mean the birth of a girl, because of the XX chromosomes. Our confidence that immaculate conception is impossible rests on limited statistical material.

I will show the relevance of these points, but what I would like to stress at this moment is that generally we may expect something new scientifically only in new circumstances. One such new condition might be another time in our evolutionary cosmology. I mean an event may have occurred at an entirely different time, with different physical constants. Under the general theory of relativity, the gravitational constant is indeed a constant, but there are schemes in which the gravitational constant does change with time. We know from experience that the relative change in the gravitational constant is less than 10^{-10} per year—an amount that is altogether negligible in speaking of civilizations that are 1000 or 10,000 light years away.

The next point, however, is more interesting. It concerns very rare events. It is very difficult to forbid them in physics. I will give you an example from steady-state cosmology. This steady-state universe concept is consistent with the expansion of the universe; but to keep the density of the universe constant, there must be a birth of new matter. But how much new matter? Only 10^{-46} grams per cubic centimeter per second. This means one atom of hydrogen a year in a cubic kilometer. A cubic kilometer at ordinary atmospheric pressure would contain more than 10^{34} hydrogen molecules. Can we, in terms of the physics we know, be confident that this is impossible, this spontaneous appearance of matter? It violates baryon charge conservation; I haven't given it very much thought, but as far as I can see, I cannot forbid such small violations of baryon charge conservation. There have been attempts to shut down steady-state cosmology. I personally think this has now been accomplished, but as a result of astronomical observations and not by reference to baryon charge conservation. What this adds up to is that it is very difficult, if not altogether impossible, to forbid very rare events.

Such very rare events may have a direct relevance to the existence of life, and this relates to the discussion of f_1 earlier. If life on our planet is indeed unique this may be due to some incredibly rare fluctuation which may involve not only the observance but also the nonobservance of physical law.

Of course, we must always speak of the physical laws we know. These suffer from limitations, and the first example that comes to mind is the interior of the neutron stars. We know the equations of state of matter down to densities of 3×10^{14} grams per cubic centimeter, the density of the atomic nucleus. At yet higher densities, the equation of state is unknown and something unexpected may occur here. I personally think everything is all right there, but I believe that at densities of a cosmological order, near the singularities which occur in cosmology, the physics we know does not operate. What is needed is a quantum theory of gravitation, which is not sufficiently developed at the present time.

If we are dealing with ordinary chemistry, with molecular civilizations, we can confidently proceed from the physical laws that we know, with the qualification about very rare events I mentioned earlier, and that applies to events on our earth in the same manner.

A few more words about two suggestions concerning entirely different civilizations at entirely different levels: First, there is a paper by Cocconi, which I have not seen but have heard about. He asks whether there might not possibly be some form of life and civilization at the level of fundamental particles. This is not absurd. We now know two hundred particles which used to be called elementary. This is far more than the number of building blocks that we usually deal with in ordinary molecular physics. Therefore, in principle, it is quite possible to obtain complicated systems from such blocks and they would have the enormous number of degrees of freedom essential for life. But that, of course, is pure speculation. I can add nothing to it.

My second example: In the *Annals of Physics* (*59* (1970): 111) there is a paper by M. A. Markov in which it is proposed that elementary particles are closed universes. I fear there is not enough time to go into this but I personally treat it quite seriously, in the sense that it is science and not nonsense. It is a very interesting theoretical possibility, but of course also quite speculative.

So, leaving aside these entirely different possibilities, I repeat once again that at the molecular level it is best to forget the possibility of wrong physics and concern ourselves with the physics we know.

My last remark is about the physics we know, which actually places very few limitations on unusual events and circumstances. Take the case of high-temperature superconductivity. This is my own field and I cannot be impartial, but this may be altogether germane to the discussion. At the present time, the superconductor with the highest known critical temperature ($21\,°K$, slightly above the boiling point of liquid hydrogen), is a certain alloy of niobium, aluminum, and germanium. The whole problem of high temperature superconductivity boils down to this: Why is it not possible to obtain a superconductor at, say, the temperature of liquid air if not room temperature? I want to claim that what we now know in physics does not in any way preclude the possibility of reaching these $300\,°K$ superconductors. Such high-temperature superconductivity may be ob-

served particularly favorably in laminated or threadlike structures, which are particularly interesting to biology. It is very
hard for physicists to prove that this is now possible, and work
is now going on very actively in this field. I therefore consider
it not science fiction to assume that evolution on some other
planet, evolution by the methods and materials we know, has
given rise to superconducting organisms. We have heard about
the likelihood of the existence of life elsewhere, but I believe
this is on the same footing as the other problems mentioned
here.

A negation of the possibility of some other laws of physics
being valid for a molecular civilization does indeed impose certain constraints upon our boundless imaginations, but it does
leave room for an enormous number of possibilities. Physics as
known today is not a straitjacket.

GOLD: One comment on the universality of the physical laws:
Astronomers have been quite proud of themselves for having
observed that on very many stars, and even in very distant galaxies, spectral lines are seen with the same patterns as they
occur on Earth, and one therefore has a strong indication that
indeed in fine detail the same physical laws apply there. Did you
have any particular reason for not invoking this as the strongest
proof of the universality of the physical laws?

GINZBURG: My whole point was that the research we have done
is on a very small subject of the universe, and the extrapolation
we are making is enormous; nevertheless I believe this extrapolation to be correct.

CRICK: I would like to make a comment on the suggestion that
fundamental particles might be used to make a new chemistry.
As Doctor Ginzburg said, we are well aware that there are a
reasonable number of building blocks; but one of the essential
characteristics for life, and especially higher life, is the evolution
of a high degree of complexity. This is done on Earth and I

think would have to be done elsewhere by combinatorial methods. This implies that you must be able to make combinations of your fundamental particles in very large variety. Therefore, if one seriously entertained this suggestion, one would have to ask whether such combinations of fundamental particles are at all likely.

There is another consideration concerning the time for which such combinations would have to last. In the system which we know on Earth, the basic time scale of chemical reactions is, say, 10^{-11} seconds. This is the pulse rate at which chemical reactions take place. You could be a bit more conservative and look at the rate at which an actual chemical reaction takes place and this might get you up to 10^{-3} seconds. But the combinations must last long enough, I should think, in order to evolve by any system of natural selection. In the system we know on Earth, the shortest generation time is more like, say, 10^3 seconds.

So, I think it is not only necessary to be able to form an extremely large number of different combinations, but the times for which the combinations last must greatly exceed the time in which the reactions take place. If a physicist working in fundamental particles could make plausible that one could have both the combinations and that they would last for a sufficient time, then I think we should take the suggestion very seriously. But is there any sign of this in what we know about fundamental particles?

GINZBURG: I did not intend to take up the cudgels for this idea. I mentioned it merely for the sake of surveying the subject comprehensively. I did so particularly because this was Cocconi's idea, who shares with Doctor Morrison the honor of being a pioneer in this field of CETI.

It seems to me that the complexity on the one hand, and the conditions on reaction rates and stability on the other, can be achieved in principle.

Doctor Gold has just suggested to me that most of the new elementary particles are unstable. But we know in a neutron star that the unstable neutrons are stable; in the same way other particles can be stabilized and, therefore, there is no objection.

CRICK: Yes, but it is not only the building blocks that may be unstable; the large number of combinations of building blocks must also be stable.

GINZBURG: It is quite possible. With 200 elementary particles you can make many more stable combinations, in principle, than in molecular chemistry.

DRAKE: Since we have gotten into the subject of macronuclear life, which has been a source of intellectual stimulation and amusement to myself and my colleagues at Cornell, I think we should add to the two requirements mentioned a third important requirement, which may be difficult to realize in the outer layers of a collapsed object, and that is a source of free energy to make life processes possible.

As was mentioned earlier by Doctor Sagan, in the case of life on Earth we have the temperature difference between the temperature of the surface of the sun and the temperature of the surface of the earth to run the heat engine or provide the free energy for life. But when we consider macronuclear life we consider it to exist in the outer layers of, say, a neutron star where the only reasonable temperature difference is that due to the temperature gradient across the dimension of the hypothetical organism, which would be very small. Some other abode for such life seems called for.

SAGAN: I would like to return to the question of possible new or alternative laws of physics. Doctor Ginzburg has suggested that there may be some unique places in which the laws of physics are different—collapsed objects, for example.

There is another possibility which he also alluded to but semi-dismissed: that there are new laws of nature to be found even under familiar circumstances. I think it is a kind of intellectual chauvinism to assume that all the laws of physics have been discovered by the year of our meeting. Had we held this meeting twenty or forty or eighty years ago, we would perhaps have erroneously drawn the same conclusion. The reason this is important is that if we imagine civilizations substantially in advance of ourselves, they may have discovered hypothetical new laws of physics about which we can only dimly guess. It may be that, for example, the preferred CETI communications channels for Type II or Type III civilizations lie within the realm of these yet undiscovered laws of physics.

Let me give an example. There are strong causality objections to the existence of tachyons, which are particles of imaginary mass which never travel as slow as the speed of light. But these hypothetical particles have at least the respectability of having been searched for in physics laboratories. If there are such particles, they are clearly the preferred way to communicate, because we are then not bound by the light travel times, which constrain all the communication channels that we usually talk about. And, as Doctor Harwit speculates in Appendix G, tachyon data transmission rates might just conceivably be enormous. I don't insist that this particular idea need be right, but it is characteristic of the problem as I see it; the solution to the CETI question may lie in laws of physics yet undiscovered.

This does not, fortunately, have very many operational consequences. We can only do our best with what we have, at whatever point in our technological evolution we are. More advanced societies will be able to guess how backward we are and will, if they wish to communicate with us, make allowances. But there will always be the possibility of yet undiscovered laws of physics; I think it would be a mistake not to bear this possibility in mind.

GINZBURG: Yes, this is a very important question. I had this at the back of my mind and shall try to confine myself to an example. Such a debate took place on a matter we are all now familiar with in quantum mechanics. In terms of quantum mechanics, we know only the probability of an electron striking a screen behind an obstacle. Dissatisfied with this, people sometimes said that some future theory, involving "hidden variables," would supply an answer other than a probabilistic one, and we would know about the motion of the electron. The argument was roughly the same as yours.

Science of course never ends. There will always be new laws and clarifications. When we say some law of physics is valid, we always bear in mind that it is true within certain limits of applicability. For example, nonrelativistic quantum mechanics is confined by a number of constraints. But this does not mean at all that future clarifications will give us an answer to the question I mention, on electron diffraction.

For molecular physics, in the phenomena that we think about when we speak of molecules, it is all within the limits of applicability of the physics we know. I see the possibility of only slight departures which, in probability terminology, are those very rare events that I referred to.

Of course, ruling out completely your example of the tachyons would be very difficult. I personally believe that there are no such tachyons and that they contradict the laws we know. But of course to have such possibilities in the backs of our minds is necessary. I spoke of the gigantic extrapolation we are undertaking, because we have factually very little to go on and we are generalizing this little to cover everything. I would describe the tachyons as an example of the difficulty of extrapolating so far.

I believe that the only rational approach today to the CETI problem is to assume a molecular civilization. If we want to use our imaginations, my preference would be to follow Markov.

All the difficulties mentioned here by Doctor Crick and Doctor Drake are then relegated to the background. We must consider a giant closed world which from the outside is seen as a tiny particle, possibly even an elementary particle. This is fantastic but it is very interesting indeed.

GOLD: There may be another point, not only that there may be different physical laws that we have not yet discovered, as Sagan discussed, but also that there are methods which are far beyond our dreams—methods which, however, are well within our present physical laws: for example, a communication channel to a distant place which uses in some way a cumulative effective energy resident in interstellar space.

MORRISON: When I first went to an astronomical meeting about thirty-five or forty years ago, I was enormously impressed by the great age of the astronomers attending, one of whom was Henry Norris Russell, who appeared to me as old as a cliff. Maybe in this context, it is not so hard for me to appear also very old and geological; but I would like to remark that the original proposals which were made a dozen years ago by Drake, Struve, Cocconi, and myself, were based on a very different principle from the ones being discussed just now, and I think that earlier principle still logically holds.

How do we differ in this discussion today from fifty years ago? While I find it hard to credit, my reading of history convinces me that the people then were as ingenious as my colleagues of today. Moreover, the idea of a plurality of worlds is at least as old as Buddhist philosophy. The difference is that, beginning a decade or so ago, we did exhibit a means of communicating over a modest interstellar distance to a society identical with our own. The difference is that a conference ignorant of coherent radiations would have had to give up. It is only radio, and possibly other coherent forms of energy transmission, that make a difference to us.

I still think the game is to begin with this singular and obviously impossible point in mind—an identical community at some distance—and to examine as cogently as we can how to spread the necessary parameters around this singular point to come to some conclusions about how far and how difficult communication will be.

The more imaginative we are, the easier it gets, but it is very hard to conclude much about beings on quasars, inside neutron stars, etc. In no way do these more imaginative suggestions make our problem harder, only easier. The sum of positive terms is always increased by adding a positive term. I think discussing points in a phase space very far removed from us is not the most efficient use of our ingenuity.

VON HOERNER: It might well be that higher civilizations have means of communication that are completely impossible for us, and maybe even completely un-understandable by us. But if they have an interest in talking with us they would know how to do it. Still, they might have a lower limit, a standard below which they are not interested, so the question then would be: Have we already reached that lower standard?

There is only one way of finding out and that is by trying.

GOLD: But I am not really willing to accept your premise, because it may well be that the means of communication they have are of a kind that we do not know how to receive, and that they would not have the means of communicating with sufficiently powerful radio or optical signals. That is something which, technologically, is too difficult for them but they would have some other means we would not recognize.

VON HOERNER: Then, we have not reached the standard.

GOLD: But they might still be interested in talking to us.

TOWNES: I would like to add just one comment to this interesting discussion on Professor Sagan's remark that the possibility of unknown physical laws has few operational consequences. Both unknown physical laws and unknown technology, I think, should have a real effect on our own procedures. That is, if we recognize some probability of these unknowns it sets a kind of limit on the efforts we should be willing to apply with those methods we now know. This, of course, has a very real significance in terms of financing.

SAGAN: There is one further way in which the existence of advanced technologies and still undiscovered laws of nature influence our problem, and that involves the possibility of a communications horizon. That is, if for convenience we imagine the evolution of civilizations to be somehow linear, so that we can talk of civilizations so many years ahead of us or behind us, then I can imagine that there is a time in our future when civilizations are so far in advance of us that their communications techniques are at least very largely inaccessible to us. It may be that we are very much like the inhabitants of, let's say, isolated valleys in New Guinea, who communicate with their neighbors by runner and drum, and who are completely unaware of a vast international radio and cable traffic over them, around them, and through them.

If we imagine that such a communications horizon exists, let us say one thousand years in our future, this means that we must exclude from communications, with an exception I will mention in a moment, all those civilizations more than a thousand years in our future. From the discussion of N (p. 166), we talked about a value of L, the mean lifetime of technical civilizations in the Galaxy, of 10^7 years, and consequently of $N = 10^6$ such civilizations. If there is a communications horizon 10^3 years in our future, then the number of accessible technical civilizations is down by a factor of 10^4. This means that instead of 1 million accessible civilizations in the Galaxy, there are only

100; and that instead of the nearest one being a few hundred light years away the nearest one is many thousands of light years away (see Appendix C).

There may be civilizations much closer to us but which are so smart we can't detect them. The ones that are dumb enough for us to detect may be so far away as to imply a very major effort in detecting them. Were we to take this consideration seriously, it would mean constructing larger arrays of radio telescopes than if we thought everyone continued to talk radio far into our technological future.

The one exception to this is the possibility that some very advanced civilizations may wish to have an antiquarian interest in communicating by obsolete communication channels, for fun or for benign interest. That may be the case, but I would imagine that it might not be the general case. The principal consequence of this alternative proposal is that the best search mode is to examine other galaxies for the few very advanced civilizations there rather than to examine our immediate neighborhood for civilizations nearly as dumb as we.

GOLD: We have an occasional interest in lower animals, and that may be enough for them.

SAGAN: It depends on how much lower. We try to communicate with dolphins and dogs and horses, but very few of us try to communicate with ants and protozoa and bacteria.

GOLD: But we study them.

PLATT: We try to communicate with babies.

SAGAN: Yes, but it is a question of what intellectual difference one imagines between us and beings millions of years in our future. If you imagine the difference to be fairly small, then there is considerable room for contact. If you imagine the difference to be immense, then I maintain there is less room for contact.

PLATT: There is the question whether they will make the effort because we are babies.

MINSKY: We have already had several communication horizons with electromagnetic radiation alone. Forty years ago or fifty years ago, the only reliable modulation was interrupted continuous wave. Then amplitude modulation became practical. But if you had a 1940s radio receiver you would have difficulty demodulating the single sideband transmission of today. Likewise with today's communication equipment, you will have difficulty decoding the shift register modulation that is becoming popular now.

So, even with radio communication, a few years of technology makes the new signals incomprehensible to the senders of the old signals.

As a civilization becomes more efficient, the transmitted signals look more and more like noise, as demonstrated in Shannon's statistical theory of coding. The most efficient communications are indiscernible from pure Gaussian noise unless you know the code word.

MORRISON: But acquisition is not signals.

GOLD: Nevertheless, somebody who wished to communicate with you might initially transmit a signal that looked just like noise.

MINSKY: He would have to make a concession of coming down to your level for some time; he would have to assume that you did not have the background to decode his more sophisticated type of modulation. But when I send a single side-band signal, I do not precede it with an amplitude signal to explain how it works.

MORRISON: As a matter of fact, we do.

MINSKY: Just by turning on my transmitting machine.

MORRISON: When you don't have a pre-arranged receiver, then you try many modes of modulation to open up the circuit. Everybody in broadcasting will tell you that, because you don't know which channel to use.

LEE: I think it is entirely reasonable to assume that a technical civilization far in advance of us would devote a small, even a tiny fraction of its resources to studying a primitive society and lower life forms, just as we devote a small but significant fraction of our resources to such enterprises. The primitive lobby on our planet are the anthropologists who would like to understand primitive societies (and the origins of our society) and the missionaries who would like to convert them. I have a feeling that most of us in this room may be of the latter variety: We would like to impart our technical civilization to other more primitive civilizations. For the sake of completeness and keeping some tidy housekeeping in the universe, advanced civilizations will devote a fraction of their efforts to contacting primitive societies such as us.

GOLD: I still fear that, although they might wish to communicate with us, their only available means of communication limits them. We may not be able to receive their transmissions. It may well be that an advanced civilization does not go in for astro-engineering, for construction on a gigantic scale, but, instead, goes in for some very sophisticated small devices. Radio might not allow them to make contact with us, while some subtler means, not yet discovered, would perhaps allow such communication.

ORGEL: I would like to emphasize that any civilization that has reached Type II in its physical technology would almost surely have reached Type II in its understanding of biology; thus, any advanced civilization which wished to communicate with us we may assume would have a knowledge of those factors which we find so difficult to estimate. They would know what sorts of biological systems, at least what sorts of Type I biological systems, they should be looking for. This suggests to me that we should not think of curiosity as to the mere existence of life as a likely motive for their interest in us.

However, I think there is at least a chance that there will be a mass of specialists, people who will want to compare civilizations as anthropologists wish to compare civilizations here. This suggests to me that if we are going to be observed, and if these civilizations wish to communicate with us, they will make it their business to use methods which we are able to detect.

If one asked what people nowadays are using antiquated modes of communication, they would be radio amateurs and Boy Scouts. It seems to me that we should watch out that it is neither of these classes that we contact when we meet our first Type II civilization.

SAGAN: I entirely agree that the subset of advanced civilizations interested in antique communications modes would make an attempt to impedance match as well as possible with our society. But how precise does the matching have to be? Might communication with us require such a precise matching that they could not possibly have the prerequisite knowledge? For example, would it be in the same class of difficulty as our having to know, let's say, Japanese? Or is it possible that there is a wide range of techniques accessible to them and understandable to us?

KARDASHEV: It is inevitable that research on the CETI problem will be conducted in several different directions simultaneously. There are many additional difficulties. We must try to establish contact with other civilizations, but such civilizations may be very short-lived. If on the other hand there do exist some long-lived supercivilizations, these civilizations may be silent. In view of the fact that we do not know which of these circumstances is likely, we must consider many different options.

MARX: I am interested in a personal meeting among representatives of Type I civilizations and in the possibility of their arrival at, say, Byurakan. We will take into account all possibilities not forbidden by the laws of nature we know.

The possible objectives of interstellar space flight are many light years away. To arrive at the next planetary system in a man's life span a speed is necessary which is near to that of light. For example, a speed of 99 percent of the light velocity c ($v = 0.99$, where v is the speed of the vehicle measured in light velocity units) is equivalent to a specific kinetic energy of 443×10^{20} ergs per gram, according to the relativistic formula,

$$B = \frac{1}{\sqrt{1 - v^2}} - 1. \tag{13}$$

For a space ship with a mass $M = 225$ tons, the kinetic energy to be concentrated onto the vehicle amounts to 10^{30} ergs. The work needed for slowing down and for returning must be taken into account as well. This energy is so tremendous that a realization of interstellar flight is possible only if the kinetic energy obtained from the fuel can be concentrated onto the vehicle with an efficiency not very far from 100 percent. Let us define the mechanical efficiency as*

$$\eta = \frac{\text{final kinetic energy of the vehicle}}{\text{total kinetic energy obtained from the fuel}}. \tag{14}$$

I calculate here this mechanical efficiency for different propulsion systems. The fundamental laws of physics (conservation of energy, momentum, and baryon number) have been used. No further restrictions on the technological possibilities have been assumed.

We first consider rocket propulsion. At an instant of time the rest mass of the rocket is m, its speed is v. Matter leaves the rocket exhaust with relative velocity w. The speed of the exhaust material in the galactic frame is u. The mass of the rocket decreases to $m + dm$ ($dm < 0$), and a fraction $\epsilon \mid dm \mid$ of the fuel

* G. Marx, The Mechanical Efficiency of Interstellar Vehicles. *Astronautica Acta* 9 (1963): 131–139.

rest mass is converted into kinetic energy. The rocket is accelerated to the speed $v + dv$ by the recoil of the burned and exhausted fuel of mass $(1 - \epsilon) \mid dm \mid$. The energy and momentum
equations of relativistic mechanics give*

$$\frac{mc^2}{\sqrt{1 - v^2}} = \frac{(m + dm)c^2}{\sqrt{1 - (v + dv)^2}} - \frac{(1 - \epsilon)dm\ c^2}{\sqrt{1 - u^2}}, \tag{15}$$

$$\frac{mcv}{\sqrt{1 - v^2}} = \frac{(m + dm)c(v + dv)}{\sqrt{1 - (v + dv)^2}} - \frac{(1 - \epsilon)dm\ cu}{\sqrt{1 - u^2}}. \tag{16}$$

Introducing $x = \log_e m$ as independent variable we derive the
differential equations

$$\frac{1}{\sqrt{1 - v^2}} + \frac{d}{dx}\frac{1}{\sqrt{1 - v^2}} = \frac{1 - \epsilon}{\sqrt{1 - u^2}}, \tag{15a}$$

$$\frac{v}{\sqrt{1 - v^2}} + \frac{d}{dx}\frac{v}{\sqrt{1 - v^2}} = \frac{1 - \epsilon}{\sqrt{1 - u^2}}\ u. \tag{16a}$$

By eliminating u and by introducing the dependent variable B
one gets

$$\frac{dB}{dx} = w\sqrt{B^2 + 2B}. \tag{17}$$

Here w, the exhaust velocity in the rest frame of the rocket, is
connected with the mass fraction ϵ by the equation†

$$1 - \epsilon = \sqrt{1 - w^2}. \tag{18}$$

The solution of the differential equation is

* G. Marx, Über Energieprobleme der interstellaren Raumfahrt.
Astronautica Acta 6 (1960): 366–372.
† It is easy to prove that the highest efficiency is achieved if the total
kinetic energy obtained from the fuel is distributed uniformly through
the ejected matter. This means that the mass factor determines the
optimal exhaust velocity w through equation (19).

$$\int_{\log_e m_0}^{\log_e m} w \, dx = \text{arc cosh} \, (B + 1) = \log_e \sqrt{\frac{1 - v}{1 + v}} \tag{19}$$

If the utilization fraction ϵ remains the same during the whole acceleration period,

$$v = \frac{1 - (m/m_0)^{2w}}{1 + (m/m_0)^{2w}}. \tag{20}$$

Here m_0 is the initial rocket mass with speed 0, and m is the final vehicle rest mass, traveling at speed v. Equation (20) was first derived by Ackeret* for $\epsilon = $ constant.

The mechanical efficiency of the rocket can be calculated easily:

$$\eta = \frac{MB}{\epsilon(M_0 - M)} = \frac{1}{\epsilon} \frac{(1 - v^2)^{-1/2} - 1}{\{(1 + v)/(1 - v)\}^{1/2w} - 1}. \tag{21}$$

In the nonrelativistic case $v < < 1$ one has

$$v = w \log_e \frac{M_0}{M}. \tag{22}$$

Consequently

$$\eta = \frac{Mv^2/2}{\epsilon(M_0 - M)} = \frac{(v/w)^2}{e^{v/w} - 1}. \tag{23}$$

Thus even an ideal rocket can work with good mechanical efficiency only in the speed region $v \simeq w$. One can conclude (see Figure 15) by saying that—

Chemical fuel ($\epsilon \leq 10^{-10}$) implies an exhaust velocity $w \leq 10^{-5}$, which can be realized by combustion engines. The working range is up to $v = 10^{-5}$. Chemical fuel is therefore suitable for flights to the nearest members of the solar system.

* J. Ackeret, Zur theorie der Raketen. Helvetia Physica Acta 19 (1946): 103–112.

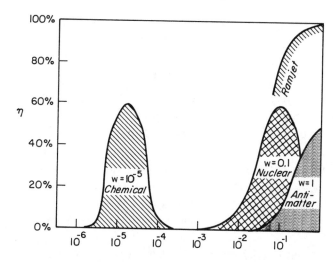

Figure 15. The mechanical efficiency η of the rocket versus velocity v (in units of the velocity of light) for four different rocket mechanisms. The parameter w is the exhaust velocity of the rocket in its rest frame. In principle the highest velocities and efficiencies are achieved with interstellar ramjets.

Nuclear fuel ($\epsilon \leq 10^{-2}$) implies an exhaust velocity $w \leq 10^{-1}$, which can be realized by electric propulsion. The working range extends up to $v = 10^{-1}$. Nuclear fuel is therefore suitable for flight to the extremities of the solar system.

Antimatter fuel ($\epsilon = 1$) implies an exhaust velocity $w = 1$, which can be realized by a photon rocket. The working range extends to $v = 1$. Antimatter/matter propulsion is therefore suitable for flight to the nearest stars.

To accelerate a ship near to the velocity of light (to $v = 1$) and to reach the stars is impossible with chemical or nuclear fuel within a human lifetime, because the mechanical efficiency of a chemical or nuclear rocket is very low in the relativistic region. But even with a photon rocket a trip to a star 100 light years away and back would need many millions of tons of antimatter.

The main difficulty of rocket propulsion is that the lion's share of the kinetic energy obtained from the fuel is taken away by the exhaust gases and only a small fraction is concentrated in the massive ship. This is a direct consequence of the momentum balance. In the relativistic region the mechanical efficiency is very inconvenient for vehicles which move freely by their own power in space. In order to realize interstellar space flight one must think about a "solid support" for the vehicle, which is able to absorb momentum without absorbing too much kinetic energy. To find a firm highway among the stars is not an easy task.

Bussard* and others have speculated about using interstellar hydrogen gas as fuel for an interstellar ramjet. Let us imagine a vehicle in continuous operation which collects interstellar hydrogen, burns it in thermonuclear reactions, and exhausts it out the back with increased velocity. In the inertial frame in which the vehicle is instantaneously at rest, the kinetic energy of the collected hydrogen plus the kinetic energy liberated in the fusion engine will be shared between the exhaust gas and the vehicle. Most of the total kinetic energy is evidently carried off by the exhaust gases. We now show that in the relativistic region the liberated nuclear energy, being only a small fraction of the available total energy, concentrates very strongly in the vehicle.† In a time interval dt a mass dm of the interstellar hydrogen is absorbed; from it $\epsilon\, dm\, c^2$ energy is liberated; the remaining rest mass $(1 - \epsilon)\, dm$ is exhausted with the velocity u, and the vehicle with constant rest mass M is accelerated from speed

* R. W. Bussard, Galactic Matter and Interstellar Flight. *Astronautica Acta* 6 (1960): 179–194.
† Cf. G. Marx, Interstellar Vehicle Propelled by Terrestrial Laser Beam. *Nature* 211 (2 July 1966): 22–23.

v to $v + dv$. The energy and momentum balance is given by the equations

$$dm\, c^2 + \frac{Mc^2}{\sqrt{1 - v^2}} = \frac{Mc^2}{\sqrt{1 - (v + dv)^2}} + dm\, c^2 \frac{1 - \epsilon}{\sqrt{1 - u^2}}, \qquad (24)$$

$$\frac{Mcv}{\sqrt{1 - v^2}} = \frac{Mc(v + dv)}{\sqrt{1 - (v + dv)^2}} - dm\, uc \frac{1 - \epsilon}{\sqrt{1 - u^2}}. \qquad (25)$$

From this it follows that

$$\frac{d}{dt} \frac{M}{\sqrt{1 - v^2}} = \frac{dm}{dt} \left(1 - \frac{1 - \epsilon}{\sqrt{1 - u^2}} \right), \qquad (24a)$$

$$\frac{d}{dt} \frac{Mv}{\sqrt{1 - v^2}} = \frac{dm}{dt} \left(u \frac{1 - \epsilon}{\sqrt{1 - u^2}} \right). \qquad (25a)$$

If the collecting surface of the vehicle is denoted by F and the interstellar gas density by ρ,

$$\frac{dm}{dt} = \rho F v c.$$

Let us denote the distance in which the mass of the collected gas equals the vehicle mass by l; and let us denote the distance variable, measured in units of l, by x.

$$ds = cv\, dt = l\, dx \quad \text{with} \quad l = M/\rho F.$$

Making use of the new independent variable x, we get

$$\frac{1 - \epsilon}{\sqrt{1 - u^2}} = 1 - \frac{d}{dx} \frac{1}{\sqrt{1 - v^2}} = 1 - \frac{dB}{dx}, \qquad (24b)$$

$$\frac{1 - \epsilon}{\sqrt{1 - u^2}} u = \frac{d}{dx} \frac{v}{\sqrt{1 - v^2}} = \frac{1}{v} \frac{dB}{dx}. \qquad (25b)$$

Eliminating the exhaust velocity u, one arrives at the differential equation

$$\left(\frac{dB}{dx}\right)^2 + 2B(B+2)\left(\frac{dB}{dx} - \epsilon + \frac{\epsilon^2}{2}\right) = 0. \tag{26}$$

A trivial solution is $B = 0$. A nontrivial solution exists if there is an initial velocity. The inverse function $x(B)$ can be computed easily from equation (26):

$$x = \int_{B_0}^{B} \frac{db}{\sqrt{b^2(b+2)^2 + (\epsilon - \epsilon^2/2)b(b+2)} - b(b+2)}. \tag{27}$$

Equation (27) contains an elliptic integral. The utilization factor of the interstellar gas is $\epsilon < 0.01$; consequently a power expansion is allowed with respect to $\alpha = \epsilon - \epsilon^2/2$. One gets

$$x = \mathfrak{F}(1 + B) - \mathfrak{F}(1 + B_0),$$

with

$$\mathfrak{F}(\xi) = \frac{1}{\alpha}\xi + \frac{1}{4}\log_e\left(\frac{\xi-1}{\xi+1}\right) + \frac{\alpha}{16}\left[\log_e\left(\frac{\xi-1}{\xi+1}\right) + \frac{2\xi}{\xi^2-1}\right]$$
$$+ \frac{\alpha^2}{64}\left[3\log_e\left(\frac{\xi-1}{\xi+1}\right) + \frac{6\xi}{\xi^2-1} - \frac{2(\xi^2+1)}{(\xi^2-1)^2}\right] + \cdots. \tag{28}$$

For relativistic velocities ($B \gg 1$) formula (28) converges very fast. The mechanical efficiency is given by the equation

$$\eta = \frac{B - B_0}{\epsilon x}$$

$$= 1 - \frac{\epsilon}{2}\left[1 + \frac{1}{2}\frac{\log_e\left(\frac{B}{B+2}\right) - \log_e\left(\frac{B_0}{B_0+2}\right)}{B - B_0}\right] + \cdots,$$

where terms of the order ϵ^2, $\epsilon^2\log_e B$, $\epsilon^{n+1}B^{-n}$ ($n \geq 1$) have been neglected. It can be seen that in the relativistic domain ($\epsilon \ll B$) an interstellar ramjet can reach an efficiency of 99 percent by nuclear fusion (Figure 15).

Another possible way to prevent the wastage of energy is to transfer the recoil momentum to the earth.* Because of the large mass of the earth, the transferred recoil energy is negligible, as in the case of a car running on a solid road. As a realization of this principle, let us consider the following propulsion system†: a departing vehicle has an effective cross-sectional area F, as seen from the earth, and acts as a perfect mirror in the backward direction. A light beam of constant intensity I coming from a terrestrial light source is reflected by it. If the vehicle moves with a speed v, the intensity I of the reflected light turns out to be smaller than I. The vehicle is accelerated to the speed $v + dv$ during the time dt. The energy and momentum balance is now

$$IF\, dt + \frac{Mc^2}{\sqrt{1 - v^2}} = \frac{Mc^2}{\sqrt{1 - (v + dv)^2}} + I'F\, dt, \tag{29}$$

$$\frac{IF\, dt}{c} + \frac{Mcv}{\sqrt{1 - v^2}} = \frac{Mc(v + dv)}{\sqrt{1 - (v + dv)^2}} - \frac{I'F\, dt}{c}, \tag{30}$$

which can be written also in the following form:

$$I - I' = \frac{Mc^2}{F} \frac{d}{dt} \frac{1}{\sqrt{1 - v^2}}, \tag{29a}$$

$$I + I' = \frac{Mc^2}{F} \frac{d}{dt} \frac{v}{\sqrt{1 - v^2}}, \tag{30a}$$

or

$$I = \frac{Mc^2}{2F} \frac{d}{dt} \sqrt{\frac{1 + v}{1 - v}}, \qquad I' = -\frac{Mc^2}{2F} \frac{d}{dt} \sqrt{\frac{1 - v}{1 + v}}; \tag{31}$$

* This idea was utilized by Jules Verne. In his novel, the trip to the moon was accomplished by a projectile from a giant gun, fixed to the earth. The problem is that man cannot survive such tremendous accelerations.
† G. Marx, *Nature* 211 (2 July 1966): 22–23.

$$v = \frac{(1 + 2\tau)^2 - 1}{(1 + 2\tau)^2 + 1}. \qquad (v = 0 \text{ for } \tau = 0). \tag{32}$$

Here τ is the ratio of the light energy $W = IF\,dt$ to the rest energy Mc^2 of the vehicle:

$$\tau = \frac{F}{Mc^2} I\,dt. \tag{33}$$

Substituting (32) into (31) we calculate the "instantaneous mechanical efficiency":

$$\eta_i = \frac{dE_{kin}}{dW} = \frac{I - I'}{I} = 1 - (1 + 2\tau)^{-2}. \tag{34a}$$

If $\tau \to \infty$, then $v \to 1$ and $\eta_i \to 100$ percent: almost the whole energy of radiation falling on the mirror regressing with relativistic speed is converted into the kinetic energy of the vehicle.

The "total mechanical efficiency" can be defined as the ratio of the kinetic energy of the vehicle (starting from rest) to the radiation energy of the beam hitting it:

$$\eta_\tau = \frac{E_{kin}}{W} = \frac{1}{1 + (2\tau)^{-1}}. \tag{34b}$$

This starts from a very low value at $v \ll 1$, and rises to 100 percent when the vehicle has reached a relativistic speed. Thus "with the solid earth below one's feet" the principal difficulty of relativistic space flight disappears. A schematic representation of the three propulsion mechanisms is given in Figure 16.

The technological difficulties in these schemes should not be overlooked. A manned space ship cannot afford an arbitrarily high acceleration. The speed of light can be approached at 1 g acceleration over a distance of the order of one light year. An ideal focusing is needed to prevent wasting the radiation energy. The laser offers the maximum degree of focusing. Even in this

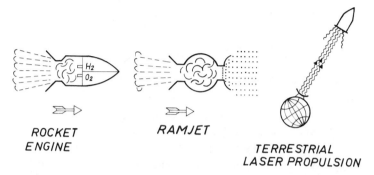

ROCKET
ENGINE

RAMJET

TERRESTRIAL
LASER PROPULSION

Figure 16. Schematic representation of three mechanisms for interstellar spaceflight: rockets (shown here as chemical, but they could equally well be nuclear or antimatter rockets); ramjets using the interstellar medium as a working fluid; and a mechanism in which the space vehicle is accelerated and decelerated by planetary laser systems.

case only x rays can give a sufficiently small wavelength to diameter ratio. The x-ray beam will be seen strongly red-shifted from the vehicle, so the mirror restrictions are not very strong.

To the present there is no evidence for interstellar visitation to the earth. This historical fact is not an argument against the existence of extraterrestrial civilizations, because the fundamental laws of physics do not allow relativistic space flight to faraway planetary systems with rockets, using an independent propulsion system. Direct contact can be realized only after having established radio contact, if at all. A possible project might be the following: The space capsule is accelerated to a fraction of light velocity by ion rocket propulsion using nuclear fuel. After this a planet-bound x-ray laser beam pushes the ship and accelerates it close to the speed of light. Ramjet propulsion will be appropriate only in passing through denser interstellar clouds. All these propulsion schemes can be achieved with considerable efficiency. Upon arrival, the deceleration of the space ship can happen in the same way, with the active assistance of the receiving civilization. The start home and the arrival at Earth follow a similar pattern.

Thus, the old science fiction idea of invasion from outer space is prohibited by the law of momentum conservation. Personal contact with an extraterrestrial civilization is not completely impossible, but it can be only a mutual undertaking of two friendly and cooperative societies.

GOLD: If you would be willing to send me on this space trip, could you assure me that the round trip would take a reasonable time? I will not accept more than 2 g acceleration for half the way to my destination, and 2 g deceleration for the other half of the way in, and the same for the return trip. Isn't the trip time well in excess of my lifetime?

MARX: If you can take 1 or 2 g acceleration and deceleration, you can get out and back within your lifetime. The main problem is that very good focusing is needed.

GOLD: The quality of the focusing of the laser beam or the matter of making a gamma ray mirror, we can ignore for the time being. The other problems are severe enough.

SAGAN: With a space vehicle capable of uniform 1 g acceleration to the midpoint of the voyage and 1 g deceleration thereafter, using the Lorentz transformation, the time to Epsilon Eridani is five years' ship time; the distance to the Pleiades is ten years' ship time; the distance to the galactic center is twenty years' ship time; and the distance to the Magellanic Clouds is twenty-three years' ship time.* In fact, you can circumnavigate the universe in your lifetime.

GOLD: That's fine, but by the time I come back I am going to find the place awfully lonely.

PEŠEK: In a short time there will be orbiting space stations with telescopes, and astronomers on board. I should like to know

* Carl Sagan, Direct Contact among Galactic Civilizations by Relativistic Interstellar Spaceflight. *Planetary and Space Science* 11 (1963): 485–498.

what are the prospects for detection of extraterrestrial planetary systems from such an orbiting station, or from a station on the moon.

GOLD: The present plans that I am aware of for the orbiting stations will not contribute significantly to the solution of this problem. The only actual planning that I am aware of which would contribute to this question at all is the proposal for very large orbiting telescopes of very large apertures which, it is proposed, will be mainly unmanned, or perhaps man-assisted on occasions. Such telescopes would then have the capability of improving the astrometric problem over what we have been able to do from the ground and, therefore, discovering around which nearby stars there may exist planets. Such telescopes would even have the possibility, in the case of the nearer stars, to observe directly by light reflection objects of the size of Jupiter. But that is still a large project and quite some way off.

The currently planned manned orbiting stations—at least the plans of which I am aware—have no such observational techniques open to them. Of course, I am not aware of Soviet plans for manned stations. Whether those include very large telescopes, I do not know.

MOROZ: Two words on this point. There is reason to think that the most advantageous wavelength for interstellar communication should be shifted into the submillimeter range, somewhere in the vicinity of one-tenth of a millimeter. If I am right, CETI, in these wavelengths, will require orbital stations since the earth's atmosphere does not transmit this wave band. But I will discuss this further later (p. 276).

GOLD: If it is true that there is any substantial advantage in using wavelengths that the atmosphere will not transmit, then of course we will want to do the job from space stations. But whether these need to be manned is quite another matter.

SAGAN [Note added in proof]: The very serious current energy problems both in quasar and in gravity wave physics can be ameliorated if we imagine these energy sources beamed in our direction. But preferential beaming in our direction makes little sense unless there is a message in these channels. A similar remark might apply to pulsars. There are a large number of other incompletely understood phenomena, from Jovian decameter bursts to the high time-resolution structure of x-ray emission which might just conceivably be due to ETI. Perhaps, in the light of Doctor Marx's presentation, we must ask if the fine structure of some fluctuating x-ray sources is due to pulsed x-ray lasers for interstellar spaceflight. But Shklovsky's principle of assuming such sources natural until proven otherwise, of course, holds. Extraterrestrial intelligence is the explanation of last resort, when all else fails.

The pulsar story clearly shows that phenomena which at first closely resemble expected manifestations of ETI may nevertheless turn out to be natural objects—although of a very bizarre sort. But even here there are interesting unexamined possibilities. Has anyone examined systematically the sequencing of pulsar amplitude and polarization nulls? One would need only a very small movable shield above a pulsar surface to modulate emission to Earth. This seems much easier than generating an entire pulsar for communications. For signaling at night it is easier to wave a blanket in front of an existing fire than to start and douse a set of fires in a pattern which communicates a desired message.

KARDASHEV: Without preempting anything that might be said by Doctor Moroz, I would like to note the possibility of detecting large astroengineering structures through their own thermal radiation and distinguishing them from dust clouds. This is a simple phenomenon, physically. A dust cloud and a large solid structure must have a continuous spectrum. However, there is a simple difference. In the case of a solid state construction, we have a purely Planckian emission. In the lower wavelength range this would be proportional to the square of the frequency. In the case of a dust cloud, the size of the dust is less than the wavelength and the dependence will be steeper than the square of the frequency.

At present, we know a fairly large number of infrared sources and yet we know only the short wave part. We need an exact study of the long wave part of the spectrum, namely, the spectrum in the 100 micron range, so as to examine what consists of small dust particles and what consists of big blocks.

MORRISON: Isn't that dependent on the thickness, the optical depth?

KARDASHEV: Yes, but in a known way.

TECHNIQUES OF CONTACT

SHKLOVSKY: We are returning now to the important problem of direct projects for establishing communication with ETI, if such exists. I must say that various participants take different views of the relative importance of the problem, the whole problem of CETI and of the problem of developing specific projects for communication. I personally believe that the general problems are no less important than the specific projects, but all of us are agreed that the present progress made by technology is sufficient to discuss such projects.

DRAKE: It has become clear in the course of this symposium that any attempts to detect ETI must be able to detect manifestations of ETI from distances of hundreds or thousands of light years. We are now turning to a discussion of the means which could lead to such detection and, hopefully, some conclusions as to which means are most promising.

As a guide to the best means, which undoubtedly is considered important by all civilizations, one can use economy in the method of detection of communication.

With this as a guide, we can immediately dismiss as very improbable various means of communication, although we must always be careful to keep in mind that we cannot predict with certainty what other civilizations might do, and so we may only produce subjective probabilities as to what is most promising.

As an example it is clearly unlikely that the means of communication is to put a note, a message on the end of a long stick and reach out to the nearest stars, even though such a stick becomes self-supporting after it is more than 22,000 miles from the earth. Similarly, as discussed by Doctor Marx, rocketry does not appear especially promising for interstellar communication.

On the other hand, as has been pointed out repeatedly, when one considers electromagnetic waves as a means of communication, they turn out to be effective, speedy, and very economical. The equation governing the range over which an electromagnetic communications link may operate is

$$R^2 = \frac{P_t G_t A_c}{4\pi P_{\text{det}}}, \tag{35}$$

where

$$P_{\text{det}} = kT_s(B/\tau)^{1/2}. \tag{36}$$

In this equation, R is the maximum range over which the link may operate; P_t is the transmitted power; G_t is the gain of the transmitting antenna or telescope. The product $P_t G_t$ gives what is known as the effective radiated power of the system. The term A_c is the collecting area of the receiving telescope, and P_{det} is the minimal detectable power of the receiving system, which we can describe to a high approximation by equation (36), where k is the Boltzmann constant, T_s is the system noise temperature, B is the bandwidth, and τ is the time constant.*

---

* Editor's note: Symbols such as B and τ have here different meanings than in Doctor Marx's previous discussion of interstellar space flight.

To illustrate how effective electromagnetic waves may be, we can take as an example in the radiofrequency range the existing Arecibo radar. Here P_t is 10^6 watts and G_t is 10^6. If we take a system temperature of $20°K$, a bandwidth of 100 hertz, and a time constant of 100 seconds, the range works out to be 2000 parsecs, or about 6000 light years, which is a very large range. In a few years this will become some 20,000 parsecs, meaning that the Arecibo radar will be visible to similar instruments throughout the galaxy.

Let us recall some history, a discussion of which leads us logically through the development of this subject. By 1960 radio telescopes had developed on earth to such a point that R had reached interstellar distances of many light years. It became sensible then to search for interstellar radio signals and this gave rise to project Ozma at the National Radio Astronomy Observatory in which radiation from the two nearest stars of solar spectral type, Tau Ceti, and Epsilon Eridani, was searched for with an 85 foot telescope. The choice of a radio frequency of 1420 megahertz was made by us for the noncontroversial and clearly correct reason that the equipment so constructed could be used for conventional radio astronomy, and no one could ever accuse us of wasting money.

At the same time, Professor Morrison and his colleague Guiseppe Cocconi at Cornell produced a more controversial, perhaps, but far more stimulating rationale for the use of that frequency, namely the suggestion that this was a unique frequency in the radio spectrum which might be utilized by a network of galactic civilizations for communication. This was, in fact, the birth of a very important concept of which we have already heard much, the idea that other civilizations may create beacon signals, special signals intended to attract and bring into contact other emerging civilizations.

In Ozma a bandwidth of 400 kilohertz was searched using a

receiver bandwidth of 100 hertz and a time constant of 60 seconds. The effective radiated power which could have been detected was 10^{13} watts, a power level which is now possible on Earth. No signals were detected, but in view of the fact that our best estimates indicate that only 1 in 10^6 stars at most might have such signals (p. 166), the results are not surprising.

Since that time, the concept of a unique frequency or magic frequencies has become a very weak one indeed in giving guidance for interstellar searches. No longer is the hydrogen line unique. We have spectral lines at the frequency of OH; we have the most intense lines in the sky from water vapor, that is, of those we know; and we have a multitude of other spectral lines. So we are now in a situation where there is no compelling argument for any particular unique frequency.

As this discouraging situation developed, another criterion for choosing frequencies was invented. That was so until the 3°K blackbody radiation was discovered and this, added to the noise system, led again to a nonunique definition of the economical frequency.

We see this in Figure 17, where we see the sources of noise, the cosmic noise at the left, the 3° blackbody radiation toward the bottom, and at the right the quantum noise, adding together to give not a T_s with a well-defined minimum but a broad minimum of constant value extending from about 1200 megahertz to 50 gigahertz. So the economical frequency is not well-defined and we are in the unhappy situation of having no compelling criteria for picking a specific frequency at which to search for beacon signals.

As another approach, Professor Oliver very early suggested an escape from this by proposing that transmissions might be pulsed so that they would exist at nearly all frequencies. This hardly lessens our problem, however, since there are as many time intervals to search as frequency intervals. Furthermore, pulses

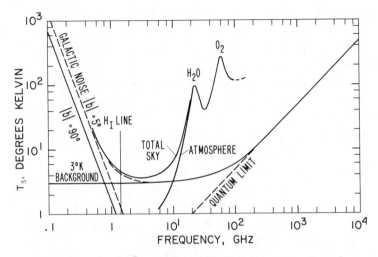

Figure 17. The system noise temperature T_s for radio astronomy as a function of frequency, allowing for quantum noise at the right. The 3°K blackbody radiation is shown toward the bottom. Galactic noise sources at two different galactic latitudes are at left.

are badly distorted by interstellar dispersion, as we know from the pulsars, so this offers no solution either. Oliver recognized this problem himself.

So at this point we reach the milestone that we could easily detect beacon signals over enormous distances. But we do not know what frequency to search on. Figure 18 shows the distance and the number of stars within that distance which can be reached using existing terrestrial equipment, all as a function of radio, infrared, or optical frequency. You see the well-defined peak caused by the effects on T_s; and I should mention that for the optical system used here, the laser power was 1 watt. Here R is the range in parsecs, and N is the number of stars within that distance. You see that we can even today reach 10^7 stars at the radiofrequencies.

It is probable that there are 1 megawatt lasers in existence

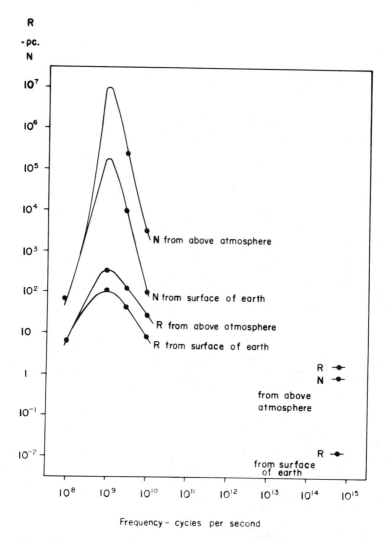

Figure 18. A comparison of various frequencies in the electromagnetic spectrum for interstellar contact. With existing equipment the distances R in parsecs and number N of stars within that distance which could with present technology be detected, are shown.

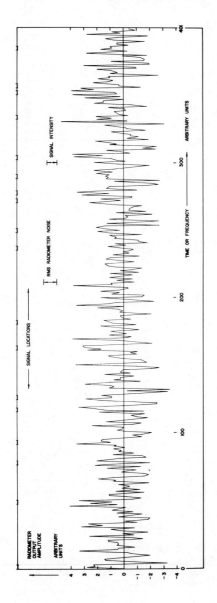

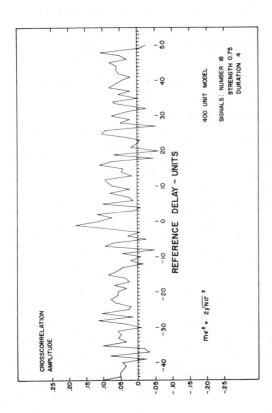

Figure 19. Above: sum of two simulated radiometer records in which 18 signals from intelligent origin of equal intensity have been inserted. Below: cross-correlation function of the two records shown above. The marked peak at 0 reference delay units indicates the presence of reproducible signals.

today which would raise the values for optical frequencies by three orders of magnitude, but they would still be four orders of magnitude less effective than the radio links.

At this point in the development of the subject, some depressing realism was faced. The question was asked: What if everyone is listening and no one is sending? This leads one to consider that it may be necessary to detect the signals other civilizations use for their own purposes—that is, to eavesdrop. In such a case, there may be many more signals, but they will come, in general, from isotropic radiation so that G_t, rather than being 10^6, is 1, losing a factor of 10^6 for us. Instead of the signals being detectable from, say, 2000 parsecs, they will be detectable from a distance of only 2 parsecs and there aren't many stars that close.

Techniques were sought which would allow us to overcome this difficulty. In Figure 19, I demonstrate one to you, a method of cross-correlating two independent spectra observed by a radio telescope from a single place in the sky. Here you see a model containing two such spectra in which there are 18 signals, weak signals inserted amongst Gaussian noise. The signal locations are marked at the top to show you where they are and you will see that it is not possible to detect the existence of this ensemble of signals from such a usual radio telescope record. However, if we cross-correlate the two records, we see a distinct peak at zero offset, such a peak showing that an ensemble of signals and, therefore, a civilization is present. Fourier analysis might be used to obtain a similar sensitivity in detecting the existence of such a civilization.

By such techniques, one may recover a factor of 100 in sensitivity, but that still leaves us down by a factor of 10^4 over what might be done with a purposeful beacon signal. So if we are to use the safe model of eavesdropping to detect civilizations, we must use very enormous receiving systems, examine the entire available radio spectrum, and use a complicated computer analysis. At this point, it is clear that a safe and effective search must look at very many frequencies with an enormous collecting area.

Is there any hope that a criterion can be developed as to where to look in the sky to speed the search? The a priori assumption would be, of course, to look at the nearest stars. However, we now understand our situation in the arrangement of the universe in a way which leads to a model in which that may not be the correct procedure. This was pointed out by Doctor Kardashev, who showed that there might, in fact, exist a population of extremely intense radio emitters from the civilizations known as Types II and III. These may be more detectable than the nearby civilizations like our own and thus there would be no reason to look at the nearest stars. It is possible to quantify this concept and I have written in Figure 20 a mathematical description for the concept. This is a slightly different version of the same result obtained by Kardashev and by Von Hoerner in that this one puts an upper limit on the power that a civilization might emit. As our discussion has suggested, this limit probably exists. By the way, this formulation is for a power law distribution of power emitters, which of course may not be a close representation of the true distribution of power.

The very striking thing which emerges is that α, the exponent, can be as great as 2.5 and still have the distant civilizations more detectable than the nearby ones. The number of civilizations can fall off as the 2.5 power, and the brightest and most distant civilizations will be seen before the nearby ones. This, by the way, is indeed the case for the radio galaxies. As an example of this formula, you see a case where $\alpha = 2.5$. Then, if for every 300 civilizations of a given power there is only one having 10 times as much power, the more distant or more powerful civilizations will be the first to be detected. This is a striking result.

The conclusion is again an unhappy one in that there is no

If $n(P)$ = No. of detectable ETI of power P in range dP and $\rho(P)$ = space density of detectable ETI of power P in range dP and S_{min} = minimum detectable flux density

Then $\boxed{n(P) = \rho(P) \dfrac{S_{min}^{-3/2}}{6\pi^{1/2}} P^{3/2}}$

If $\rho(P) = KP^{-\alpha}$, then $\dfrac{n(P_2)}{n(P_1)} = \left(\dfrac{P_2}{P_1}\right)^{3/2-\alpha}$

$$\frac{N_{P>P_1}}{N_{P<P_1}} = \frac{\displaystyle\int_{P_1}^{\gamma P_1} n(P)\,dP}{\displaystyle\int_0^{P_1} n(P)\,dP} = \boxed{\frac{\gamma^{5/2}-1}{1}}$$

>1 if $\gamma^{5/2-\alpha} > 2$

or $\boxed{\alpha < 5/2}$

Example: If $\alpha = 5/2$, $\dfrac{P_2}{P_1} = 10$, $\dfrac{\rho(P_1)}{\rho(P_2)} \approx \underline{\underline{300}}$

Figure 20. Relations connecting the distribution of extraterrestrial intelligences (ETI) showing circumstances in which the most distant civilizations may be the most readily detectable. See text for further details.

strong case for observing the nearest stars or, indeed, any particular point. Although we have the power to discover civilizations, we know neither where to look nor on what frequency.

Well, there is one happy situation and that is that we do indeed seem to know how to decode at least some forms of messages, as has been discussed in many places.

The general conclusions regarding electromagnetic radiation links are as follows: The search for reasonable beacon signals is easily and effectively carried out with existing equipment, although very much search time is required. A reasonable search in the eavesdropping mode or a search for very weak beacons requires very much larger antennas than we now have. It is reasonable to believe that a well-planned automated and lengthy search can lead to success in the search for radio signals from other civilizations. But it is clearly wishful thinking to think that such a search will succeed with limited resources and limited time.

KARDASHEV: I agree with many of the points just made by Professor Drake. Most important of all is the question of choosing the optimum strategy in our quest.

I would like to elaborate upon such a strategy and what should be done to optimize it. Let us denote by 1 the strategy of the sender and 2 the strategy of the receiver. The strategy should be organized roughly as follows: First, the energy for transmitting one bit of information should be minimized. Second, interference in the vicinity of the sending side should be minimized. Third, the cost of the receiving apparatus should be minimized. Fourth, a signal-to-noise ratio greater than unity, considering the conditions of wave propagation in the interstellar medium, is desirable. And lastly, the point I should particularly like to stress, is that the ultimate time must be minimized.

The last point need not be stipulated but it is obviously a limiting factor. Evidently it is quite true, as Professor Morrison pointed out, that it would be rational to produce a special call signal that could be singled out, signals that would be keys to deciphering and decoding the information, signals that would point to the language of the information, and signals that would contain the information itself.

Apparently the optimum strategies for each of these cases differ. We must first of all speak of the signals intended to attract our attention. Evidently these beacon signals must be tremendously redundant in terms of the Shannon relationship in information theory and must satisfy all five points I have mentioned.

It seems to me that at present the most satisfactory case that satisfies all these points is rare impulses of very great power. There is no problem of frequency search if the bandwidth $\Delta v \approx v$. The duration of the impulses and their power will be discussed presently.

What must be the model for the sending side if it wishes to transmit one bit of data by one impulse of very great redundancy? If we proceed from the present technological level on the earth, the greatest power we can count on is the power already produced under terrestrial conditions—namely, the energy of the maximum explosion produced on Earth, on the order of 10^{24} ergs. As pointed out recently by Sterling Colgate,* a powerful explosion can be transformed into electromagnetic impulses of equal power. How is this done? If you have a dipole magnetic field linked to a star or to some artificial body (Colgate examined this for a supernova), in the explosion the magnetic field undergoes a rapid deformation and the changing energy of the electromagnetic field produces the emission of a single powerful impulse. If we speak of reproducible conditions, then, in principle, we can take some external fields, such as the terrestrial or solar magnetic field; if we can set off the explosion here, the result

---

* S. Colgate and P. Noerdlinger, Coherent Emission from Expanding Supernova Shells. *Astrophysical Journal* (1971): 509–522.

will be a deformation of the magnetic field and the deformation of that magnetic field will produce a single very powerful electromagnetic impulse. The spectrum of that impulse has been calculated by Colgate and is in the radio spectrum. The spectrum depends on specific parameters of the magnetic field and the duration of the blast can be regulated.

In this way, we can regard this quantity, 10^{24} ergs, as a limit, a boundary where our technological capacity ends for transmitting one bit of data. This is the boundary for today.

Doctor Drake spoke of the formidable difficulties arising in finding the optimum radio astronomical wavelength range. I would like to add to this the difficulties involved in the uncertainty of the strategy. The first type of strategy for establishing communication could be the following: Assume there is an isotropic transmitter and a receiver which also operates isotropically. In this strategy there is no previous arrangement between the correspondents and they are unaware of one another's existence.

In a second strategy, the transmitting side operates isotropically, but we, the receiving side, for some reason have selected a solid angle we consider particularly promising. Reception in this case is beamed. This may be determined by the accuracy with which we know the coordinates of the star or the angular dimensions of the planetary system; or by the error we know to exist in these coordinates. This would also imply that we have the means for building a big radio telescope with a highly anisotropic antenna pattern.

There may also be a third and fourth case: the third, opposite to the second case, and a fourth case where both sides know each other's coordinates and direct their apparatus at each other. All these cases may be calculated by present-day communications theory and in each case we will get an optimal range or spectrum.

There are also difficulties arising from the fact that the recipient may build up his isotropic system in a different manner. For isotropic reception we may use a simple dipole antenna or we may use a gigantic number of parabolic mirrors to cover the whole sky. The cost of these two experiments will differ by several orders of magnitude. Until all these calculations are made, we must consider the background noise, the noise of the star near which the planets are. In each of these cases, there is a distinct optimal frequency range.

The most interesting case is that in which we stage a simple experiment. The sender and the recipient use the simplest antennas, dipoles, and then we can solve our problem in strict terms and obtain the approximate energy needed for transmitting a single bit.

At present, thanks to the pulsars, we have studied the propagation of radio waves in the interstellar medium quite well. Radio astronomers are quite familiar with the fact that in the propagation of a radio pulse through the interstellar medium, dispersion is produced by the medium causing an extension of the pulse in time, a broadening of its spectrum, and in some frequencies a fading of the pulse. All this affects the optimal duration of the pulse and the width of the band. We can calculate all these conditions in such a way as to assure that no pulse is lost and we can preserve all the conditions of minimum cost spoken of here earlier. In the simplest case, it appears that the decimeter wavelength minimum in the background is advantageous and such an experiment could be very inexpensive (I will give you a rough cost estimate shortly). Since this is the simplest type of experiment it is the one that should be given priority. Similar calculations can be made for more complicated cases, and in all the other cases we naturally get different results.

The background radiation of the sky has been studied in practically all parts of the spectrum except the far infrared. The far infrared may also be optimal. Its optimal character may be related to the fact that dispersion in the interstellar medium and thus the fading of the pulse and its extension in time is insubstantial there.

We can build a similar isotropic system in the form of a bolometer, in the simplest case in the form of a sphere, which measures its own temperature change due to the incident radiation. Such a system may also prove attractive for long-distance communication. The physical advantage of a bolometer lies in the fact that for isotropic radiation and a dipole antenna, the collecting area is proportional to the square of the wavelength. This being so, the short waves are disadvantageous. If we use a bolometric system, on the other hand—a sphere, say—its surface does not depend on the wavelength, and it may turn out that the short wave part will be more advantageous. Calculations show that everything depends on the infrared background. If the minimum intensity of the background in the infrared turns out to be as low as in the radio part of the spectrum, the submillimeter interval may prove to be more advantageous than the radio.

My second example concerns calculations for optimizing the strategy for observing the center of the Galaxy. Let us assume that we make an antenna with a beam corresponding to the angular diameter of the center of the Galaxy, about 2 minutes of arc. In this case, the noise determining the strategy is a spectrum of the radio emission of the center of the Galaxy and the background emission plays no role at all. In this case, if we take the real value of the flux density from the Galactic center, we find that the optimal wavelength is 5 centimeters. The minimum energy for isotropic transmission of one bit is 10^{33} ergs. These conditions can be achieved if the ground receiving antenna has an effective area of 6000 square meters. This is a radio tele-

Table 2. Parameters for Various Modes of Interstellar Radio Communication

Type of Communication	R Distance in Centimeters	Δt Pulse Width (from Dispersion) in Seconds	W Pulse Energy in Ergs	P_m Average Power in Ergs/Second	P_p Peak Power in Ergs/Second
Bracewell probe in our planetary system, λ = 21 cm, antenna gains $G_1 = G_2 = 1$	3×10^{13} (2 a.u.)	10^{-7}	10^{15} (10^8 J)	10^{10} (1 kW)	10^{22}
Directed antenna contact with nearest stars, λ = 21 cm, antenna diameters $D_1 = D_2$ = 200 m	10^9 (10 l.y.)	3×10^{-4}	10^{15} (10^8 J)	10^{10} (1 kW)	3×10^{18}
Isotropic emission from the Galactic nucleus, λ = 5 cm, G_1 = 1, receiving antenna D_2 = 100 m	3×10^{22} (3×10^4 l.y.)	1	10^{33}	10^{28}	10^{33}
Directed antenna contact with Andromeda galaxy, λ = 21 cm, $D_1 = D_2$ = 10 m	2×10^{24} (2×10^6 l.y.)	3×10^{-2}	10^{32}	10^{27}	3×10^{33}
Distant metagalaxy, isotropic emitter, λ = 21 cm, $G_1 = G_2 = 1$	10^{28} (10^{10} l.y.)	30	10^{48}	10^{42}	3×10^{46}

We adopt here a bandwidth $\Delta \nu = \nu$, a time repetition of pulses, 1 day, and a signal-to-noise ratio of 10^3.

scope dish with 100 meter diameter (Table 2). Such calculations can be made for any source—yielding different values, of course. For very compact sources which have the same flux as the Galactic center, the optimum range will again move into the short-range part of the spectrum. Parameters for a range of other communications circumstances are shown in Table 2.

Before drawing the necessary conclusions from all this, I would like to say that the conclusions regarding the optimal wavelength for transmitting the actual information can be made more definite if once again we assume a signal source unrelated to any other powerful radio source. In that case, everything will depend on the strategy adopted in the receiving apparatus. But calculations with due consideration for the fluctuations and the background show that in the two limiting cases the optimal wave band must correspond either to the minimum intensity of

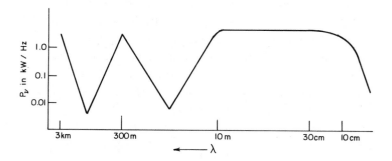

Figure 21. A rough estimate of the broad band radio noise generation on the planet Earth at the present time. The power in kilowatts per hertz is shown as the ordinate, and the wavelength increasing to the left is shown as the abscissa. The curve represents the integral of all transmitting facilities registered by the International Scientific Radio Union. Other facilities also exist.

the background, or else, in the case where we have a single large antenna, the best region would be around 2 millimeters.

Briefly about eavesdropping potentialities: I would like to show you roughly the spectrum of the noise generated on the earth at present in a broad wave band. The general pattern is shown in Figure 21. The wavelengths are plotted logarithmically, and the overall power of the transmitter is shown per hertz: The scale is also logarithmic. Shown are data for all transmitting facilities registered at Geneva by the International Scientific Radio Union. They have data indicating how the spectrum is growing with time. If we integrate all this to get the full power, it will prove to be larger than the real power at the command of the human race. This means that most of the transmitting facilities are registered but not in use. On the other hand, of course, there are facilities that are not registered. Many facilities work for a very short time.

Up to 30 centimeters wavelength, the full power per hertz exceeds the full power per hertz emitted by the sun. An external observer would be able to see a more powerful radio emission from the earth than from the quiet sun. However, this is a very small absolute quantity, and radio astronomers have not been able to observe such radio emission from the nearest stars. The antennas needed for such emission from stars within 10 light years must have an area of roughly 10^6 square meters. I think that these data show that the eavesdropping problem is quite realistic to work on.

Finally, I would like to say something about the cost of such a project in the very near future. First, the cheapest proposal—let us call it a two-island proposal. It is proposed to set up two stations with receivers of high sensitivity and dipole antennas. These points must be chosen very carefully so that they correspond to a minimum ground interference. I hope that such points can be found within the next few years. In five or ten

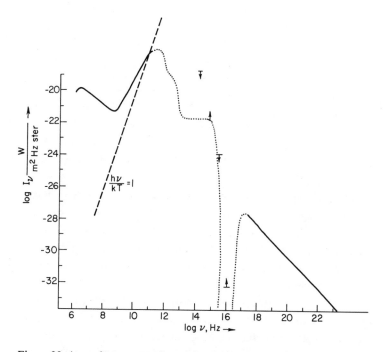

Figure 22. A rough representation of the sky background. Radio frequencies are given by the solid curve at left, optical frequencies by the solid curve at right. The broad middle range of infrared wavelengths, which for all we know has depressions in it, must await observations from above the atmosphere.

years, this will probably be impossible because the power of the emitting facilities, according to the statistics, doubles every five years. (Especially important is emission from earth satellites.)

The sort of receiving arrays to be used will be determined completely by the sky background, roughly represented also in Figure 22. A maximum in the background emission exists somewhere about 200 meters wavelength; a minimum radio background somewhere in the neighborhood of 50 centimeters; and there is an unexplored region at wavelengths shorter than 1 millimeter. It may also have deep depressions. This is a separate problem which is now being studied in astrophysics. But what can be done now is to set up an isotropic system in the region of the minimum radio background, and possibly subsequently at other frequencies. If there are two receiving sites, coherent selection will enable us to determine whether we are receiving interference or a signal from space.

This problem corresponds to an astrophysical problem. I have spoken of Colgate's paper where he predicts that in this way it may be possible to detect supernovae flares even in other galaxies. It would be ideal, of course, to cover the entire radio spectrum but that would be rather expensive. The cost of such an experiment would be about a million rubles or dollars for the two stations.

SHKLOVSKY: That would be very cheap.

KARDASHEV: I am speaking of the cost without an antenna, a dipole system. Undoubtedly one could propose far more expensive experiments calling for the construction of very large antennas. I have spoken of the fact that the detection of, say, the radio emission of our civilization from a distance of ten light years would require an antenna of 1 square kilometer in size. Such a system is highly desirable in other wave bands, too. It must evidently cost much the same as other antennas but will have a correspondingly smaller area. It is extremely desirable to build such a telescope.

PANOVKIN: A question to Kardashev. I didn't quite grasp in what sense you used the term "strategy." I observe that you are speaking of different variant cases for different objects and different modes of transmission and the matter of calculating these correctly. A question of strategy, however, is a very important one. This is one of planning the operation, how to pass from one variant to another, and such cases rule out others, making it possible to vary the planning of the operation. There is also the theoretical interest in planning an operation in terms of a game strategy for two civilizations which find one possibility unrealistic, pass on to the next, and so on.

KARDASHEV: Here, of course, I was speaking only of a strategy in terms of experiments that can be staged in the near future; that is all.

PANOVKIN: I repeat, this is not a strategy.

MORRISON: A question for Doctor Kardashev. It is indeed very attractive to hear such a clear, simple account of the problem, especially because I fully agree that we need to separate very sharply the problem of acquisition of signal from the problem of receiving an extended message. On the other hand, I would like to put the following question: Suppose we imagine a wholly foggy universe where electromagnetic radiation of all frequencies, while not absorbed, is scattered randomly—would not his proposal be exactly suited to such a case? If this is the case, as I think it is, then it seems to me that he throws away in such a proposal our extensive knowledge of the geography of the universe, even though we cannot certainly predict the best location of transmissions.

GINZBURG: In other words, what do we need a dipole for? Are you counting on an explosion?

KARDASHEV: Why is it necessary in the first experiment to use a dipole? Only because this is the simplest experiment and no one has yet performed it. At present, there are so many pulses generated on the earth that radio astronomy is too late. It has missed the boat to receive such pulses from outer space and, therefore, there is a need for a special system consisting of two or more stations operating coherently so as to select the pulses and decide whether they may be coming in from outer space or from the earth.

I do not count only on an explosion. Any type of generation is suitable, but I used an explosion as an example since this could be accomplished even by us.

DRAKE: I believe a network of stations has been operating for some years in the United Kingdom to search for similar pulses and has so far not discovered any.

Do you have any estimate of the efficiency with which the energy of a hydrogen bomb explosion is converted into electromagnetic emission by the system you have described? That efficiency is rather important.

KARDASHEV: According to Colgate's concept, several tens of percent of the explosion can be converted into a single impulse.

SAGAN: James Elliott of Cornell has prepared a short discussion on nuclear explosions and CETI for this meeting. He draws the following conclusions (they are described in more detail in Appendix H). With x-ray detectors now in use, the Starfish thermonuclear explosion, which Elliott says had a yield of 1.4 megatons, could have been detected over a distance of only 400 astronomical units in x-rays. He also asks what would happen if all the thermonuclear weapons in the stockpiles of the United States, the Soviet Union and other nations were simultaneously detonated—say in space on the far side of the Moon—for such a purpose. (Such an undertaking seems desirable quite apart from interstellar communication!) Making a guess as to what these stockpiles are and assuming that the x-ray pulse could somehow be concentrated into a conical beam of 30 degrees half-angle, he finds the distance at which the pulse can be detected is some 190 light years. If the receiving civilization were not watching at that moment that's it. It does not seem to be a very efficient method.

GOLD: I have thought from time to time about how to make a transmitter using explosives, perhaps nuclear explosives, but one which obtains a much higher efficiency than usual, a device into which you focus the explosion and obtain a continuing signal with considerably restricted frequency spectrum rather than merely a single delta function in time over all frequencies. Such a device can take the form of periodic structures in which magnetic fields exist to start with and the periodic structures are crowded out as the explosion wave runs through it. In this way one can obtain a periodic signal that runs, of course, at the speed of light. I then thought it would be best to take such devices, which could perhaps produce on the order of 10^{20} ergs in a pulse, and put them into some large thin Mylar dish in space which would of course then be destroyed by the explosion, but which would serve to direct the pulse in a desired direction.

OLIVER: Any system for attracting attention that relies on an occasional short impulse very greatly reduces the L in equation (1). For example, if we set off all the nuclear bombs in the world to attract the attention of another civilization, which might be the best use for these bombs, L would then become something on the order of a microsecond.

GINZBURG: I would like to say that, in my opinion, nuclear explosions are completely unsuitable. This has already been said, and I will not dwell upon this any further. I only want to ask Kardashev this: If you have transmitters, ordinarily powerful transmitters in civilizations of Type I, can you detect anything with a dipole? Is your dipole any good for powerful transmitters without an explosion, if we rule out explosions which are very difficult to control?

KARDSHEV: Doctor Drake spoke earlier about the possibility of using radar devices and showed that at 10 kiloparsecs it would be possible to establish contact without a dipole antenna. But it does not seem to me so terrible to use explosions for transmitting impulses. The principle is very simple: you create, say, over a period of a year a strong magnetic field without any explosion. You pass a current through a coil, you create a field, and when you need it you make a short circuit. The self-induction of the system is calculated and you get a short impulse. Such a system would be used especially to emit one impulse. It seems to me quite realistic.

GINZBURG: The dipole is just for that? I see.

TOWNES: One of the criteria which has been left out of this discussion is the desirability of two-way communication. This very much affects, it seems to me, many of these arguments about range. I would be much more interested in learning of life on a star 5 or 10 light years away, so that during my lifetime there would be some chance of communication, than in finding one 100,000 light years away. There are many techniques which allow fairly easy (at least easy on our scale) communication up to a few thousand light years and many stars within that range. I would think one ought to put rather heavy weight on reasonably cheap and constant searches of these nearby stars.

SAGAN: Any rational program should of course include nearby stars. There are two points, though, on which I would like to lay stress. The first is that one cannot simultaneously have civilizations which are close to us in space and close to us in time; this is a simple consequence of equation (1). Statistically, a civilization which is near us in space is going to be very far from us in time. That an advanced civilization lies near us in space implies that civilizations are common, and therefore that they have a long mean lifetime. Therefore any random one would be very old. I suspect that very old civilizations will not be the ones that will seek contact with us (cf. pp. 212–214).

The second point is that if you believe there are civilizations within the nearest 10 or 20 parsecs, you are taking far more optimistic values for the various f's in equation (1) than I adopted.

DRAKE: I wish to be clear on Doctor Townes' point. If $\alpha > 5/2$ (Figure 20) it matters not where you look in the sky. If $\alpha < 5/2$ only the nearby stars count. If you look at the nearby stars, you are making a search suitable to either case, but you should not fool yourself into thinking that you are increasing your probabilities of success in so doing.

VON HOERNER: The exponent on Drake's distribution function determines whether we should search the whole sky at once or for single stars only. In the first case we would want very poor resolution, in the second case higher resolution. So if we build a large array of single telescopes, in the first case it should be very compact and in the second case it should be widely scattered.

SHKLOVSKY: I think that Kardashev had this in mind, too—the preferential establishment of contact with a civilization of Type II. In such a situation, it is important to stress the expediency of using interferometric methods. If we use long baseline interferometers and in this way resolve angular differences of 10^{-4} seconds of arc, and if we consider such a civilization to have characteristic dimensions of 1 astronomical unit, it is easy to see that such a civilization is easily detectable from any point in the Galaxy. We can conceive of a system of very large mirrors scattered over a distance such as the earth's diameter. Such a system would be very effective.

MINSKY: We have not found a natural unique frequency but it should be noted that antennas are expensive and receivers are potentially extremely cheap. Perhaps, for example, if the antenna is a spectroscopic device so that it collects all of the radiation and focuses, by dispersion, different frequencies at different points, we can build one million small semiconductor receivers spaced along the scattering diagram of the telescope. It seems to me that if this were a serious project, each receiver would cost only a few dollars and with the linearity of space and with proper design, they need not interfere with one another, so that one could have an enormous array of inexpensive receivers. With suitable design, it is even conceivable that a large, long-focal-length reflector could form a small radio image.

GOLD: Has anybody contemplated that there is the possibility of a communication channel by what is commonly called the whistler mode, along the magnetic field lines in the Galaxy at very low frequencies, starting from outside the solar system where the plasma densities go down to the Galactic value? The advantage of such a communication system would be that you do not suffer inverse square attenuation, because the signal is guided along the field lines. If you are lucky enough that some other guy sits on the same field line as you do, more or less, then you have a possible extremely low attenuation signal channel.

TROITSKY: The quest for monochromatic emissions from stars in the vicinity of the sun is my first subject. I intend to discuss the quest for monochromatic emission on the 21 and 30 centimeter wavelength bands from stars at a distance of tens of light years. We examined twelve different astronomical objects, mainly stars of G spectral type. This investigation was based on the assumption that the most suitable signal, most artificial in character, is the purest sinusoidal signal.

The transmission of information by such a signal is sufficiently slow. For purposes of reception, we developed a narrow band receiver on 21 and 30 centimeters. It made it possible to conduct observations simultaneously in 25 frequencies. In a 13

Table 3. Astronomical Objects Examined for Interstellar Communication at Gorky

Object	Distance in Light Years	Stellar Spectral Type	Number of Observations
Epsilon Eridani	10.8	K2 V	6
Tau Ceti	11.9	G8 V	6
380 Ursa Majoris	14.7	dM0	2
Rho Coma Berenices	27.2	G0 V	7
Beta Canis Venaticorum	30.2	G0 V	7
Eta Bootis	31.9	G0 IV	6
Iota Persei	38.8	G0 V	4
47 Ursa Majoris	44.7	G0 V	10
Psi-5 Aurigae	48.6	G0 V	2
Pi' Ursa Majoris	50.2	G0 V	4
Eta Herculis	61.6	G8 III–IV	3
M31 (Andromeda Galaxy)	—	—	6

Source: V. S. Troitsky, A. M. Starodubtser, L. I. Gershtein, and V. L. Rakhlin, *Astronomicheskhii Zhurnal* 48 (1971): 645.

hertz band next to each wavelength, the frequencies were spaced 4 kilohertz apart. Twenty-five filters were smoothly interchanged. This made it possible to examine frequency bands of 2 megahertz. The examination of this interval took 10 minutes.

The noise temperature of the receiver in both frequencies was about 100°K. Reception was conducted by a small antenna 15 meters in diameter in Gorky. The threshold sensitivity of the system for a flux of a sinusoidal signal was about 2×10^{-22} watts per square meter.

The observations conducted so far have been only at 30 centimeter wavelength in a 2 megahertz band.

The observations were conducted in September and October of 1968. In Table 3 are listed the target objects.

We carried out sixty-five observation sessions. That was about five sessions per star at different times. We may say that with an accuracy down to 2×10^{-21} watts per square meter we observed no monochromatic radiation in the 30 centimeter wavelength. As you see, this was similar to the Ozma project carried out by Drake, although the instruments differed substantially. We used a nonradiometric receiver—an ordinary resonance receiver.

My second communication concerns the quest for sporadic emissions which could be a result of ETI activity; in other words, whether a sufficiently powerful pulse, a pulse coming from outer space could be detected. One of the ways of detecting ETI is a quest for the results of its engineering activity. One such result could be electromagnetic radiation. We may assume that this emission may be variable in time. Possibly there may be a sporadic rise in the emission, a rise of different duration, or there may appear impulses of short or long duration—the duration may vary.

In accordance with this concept, we conducted observations over the entire decimeter band. We used a radiometer of 600°K noise temperature in the 50, 30, and 16 centimeters, with a dipole and reflector. The forward lobe of the dipole antenna pattern was pointed at the zenith. To distinguish the signal from space signals of local origin, the observations were conducted at several locales simultaneously. The measurements were conducted from March 1970 between points spread out over 1500 kilometers between Gorky and the Crimea. A recording instrument was used; in one minute the tape moved at such a speed that we could record simultaneously events with an accuracy down to one second.

The very first measurements at 50 centimeters showed coincident phenomena, occurring mostly in daylight. It became obvious that statistical data could only be observed as a result of prolonged observations and the nature of the phenomena could only be established by covering large areas in longitude and latitude.

Accordingly, in 50, 30, and 16 centimeters, we organized simultaneous observations at four points, spaced meridionally over
distances of about 1500 kilometers, and separated by 8000
kilometers in latitude. The instruments were placed in Gorky,
the Crimea, Murmansk, and the Ussuri region of Siberia.

The observations were conducted for two months from September 1 to November 12 of 1970. The conditions were two
days' observations, two days' intermission. We observed phenomena which may tentatively be broken down into four groups.
The first group are singular blips tens and hundreds of seconds
in duration, and they were observed against the background of
instrument fluctuations. I show you an example of this in Figure
23. Here time is plotted on the abscissa and the antenna temperature on the ordinate. There are three recordings: from the

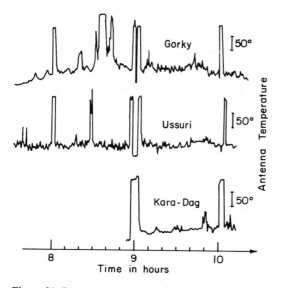

Figure 23. Sample run in daytime of the cross-correlation of signals received by isotropic detectors in Gorky, in Siberia (Ussuri), and in the
Crimea (Kara-Dag). Note the correlation of events.

Crimea, from Gorky, and from Ussuri. You see several coincidental events. We have a number of similar examples. That is the first character of the impulses observed.

The second group of phenomena is the increase in fluctuation resembling a noise storm of different intensity. The third group are the blips on the background of a noise storm. This is a periodic type of impulse. Finally, we observed a fourth group, a comparatively long-lived change in the level of emission unaccompanied by noticeable, appreciable fluctuations.

The processing consisted in determining the statistical characteristics at each point and in finding coincidental characteristics. It was found that the observed coincidental parts were far higher than the chance figure, but this was observed mainly in daylight. At night, the coincidences were fewer and approached the random figures.

The existence of these coincidental events, the number of coincidences exceeding random values, indicates that there is some global reason for these phenomena. However, an analysis of the shape of these coincidental curves and their association with daylight suggests that the radio emissions in question do not come from outer space but apparently arise in the atmosphere and may be traced, in the final analysis, to solar activity. Therefore, the experiment did not detect any sporadic emissions reaching us from the Galaxy.

The experience gained in this investigation suggests that it is most expedient to seek sporadic galactic emission on shorter wavelengths, decimeter wavelengths and centimeter wavelengths, probably in the 5 to 15 centimeter range where the camouflaging or disguising effect of natural earth sources is sufficiently low.

SAGAN: Professor Troitsky is to be warmly congratulated on the report that he has presented to us. It is the first attempt to do an Ozma-type experiment since Ozma, and the first one reported in about a decade.

SAGAN (Note Added in Proof): Since this CETI Symposium two further efforts to detect radio signals from extraterrestrial civilizations have surfaced, performed with the 140-foot and 300-foot aperture telescopes of the National Radio Astronomy Observatory (NRAO), Green Bank, West Virginia—where Project Ozma was initiated. Both enterprises use the 21-cm hydrogen line and are not restricted to sinusoidal signals. The first effort, by G. Verschuur, examined ten nearby stars (see Table 4). The results are published in *Icarus* 19 (1973): 329–340. The second effort, conducted by B. Zuckerman of the University of Maryland and P. Palmer of the University of Chicago, is still in progress (May 1973) and will have examined ~ 200 stars when it is finished. Both efforts have yielded negative results. This is hardly surprising. According to the previous discussion (p. 166), even with parameters that some consider optimistic we must examine $\sim 10^6$ stars to have a fair chance of coming upon a single extraterrestrial civilization. Thus the probability of suc-

Table 4. Stars Examined for Interstellar Communication at NRAO

Star	Distance in Light Years	Stellar Spectral Type
Barnard's Star	6.0	M5
Wolf 359	7.8	M8
Luyten 726-8	7.9	M6
Lalande 21185	8.2	M2
Ross 154	9.3	M6
Ross 248	10.3	M6
Epsilon Eridani	10.8	K2
61 Cygni A, B	11.1	K3, K5
Tau Ceti	11.9	G8
70 Ophiuchi A, B	16.4	K1, K5

Source: G. L. Verschuur, *Icarus* 19 (1973):329-340.

cess of all efforts to date is $\leq 10^{-4}$. Nevertheless, the use of radioastronomical facilities now in existence or under construction can increase this probability by many orders of magnitude.
GINDILIS: I will begin by dealing with some general principles of a CETI system, the strategy of CETI, and then I will pass on to one possible system of beacon signals using impulse signals with compression in the interstellar medium.

To be more definite, let us assume we are using an electromagnetic channel. The specific character of a CETI system is such that no correspondent knows in advance what the other is up to and he can only assume what strategy the other follows; on this basis he seeks to coordinate his actions with the actions of the correspondent.

For example, the recipient may make certain assumptions about the system of transmission employed by the sender and on the basis of these assumptions he will use a certain mode of reception. In turn, the sender must take into account the methods of reception that are to be used by the recipient and he must do so on the basis of his assumptions concerning his sender's actions. What this amounts to is a game situation, a typical game situation. The specific feature of this interstellar communication game—unlike a game between the communications experts of hostile armies—is that the correspondents, instead of seeking to upset one another's schemes, are trying jointly to find a solution to the problem that will enable them to make the game a success.

A solution of this problem is facilitated by the fact that there are certain common elements, designed neither by the sender nor the recipient. Such an element is the communication link itself. By this link I mean the region of cosmic space between the sender and the recipient, between their antennas—the interstellar and interplanetary medium, and the planetary atmospheres.

A study of the parameters of this link makes it possible to draw certain conclusions as to how such a system should be arranged, or at any rate as to how it should not be arranged. For one thing, we may make definite conclusions about the optimal wavelength in CETI. This being so, we are guided by certain objective laws and are endeavoring, on the basis of such laws, to formulate certain rules of the game. I think that such objective rules of the game should be supplemented by one subjective rule and that may be called the principle of least uncertainty.

According to such a principle, every participant in the game, in his assumptions concerning the actions of the other party (and on the basis of certain practical conclusions), bases himself on the rule that the uncertainty must be reduced to a minimum and that, therefore, it is necessary to take into account only factors of principle.

I will not go into this at length, but I will say that in developing a CETI system it is natural to single out two tasks (this has already been said) : the transmission and reception of the beacon signals, and the information communication. I will speak only of the call signals.

Several assumptions may be made concerning the character of the call signals and the requirements that must be satisfied by a system of communication for their transmission and reception.

First, the call signals are intended to facilitate the task of detection; specifically, they must facilitate the establishment of the artificial character of the source. For this purpose, it is necessary, in addition to certain physical characteristics of the signal, such as the spectrum and the statistical structure of the signal, that there also be a certain amount of semantic information contained. I will not discuss the character of that semantic information and will confine myself to discussing its quantity.

On the basis of certain considerations which I will not go into

here, I postulate that the amount of information in the CETI
call signals is not very great. Hence, it follows that the transmitting capacity of channels for transmission and reception of
signals is not of crucial importance.

Secondly, the coding scheme must be as simple as possible.
These probably are the most important features of the signals,
for otherwise they will not serve their purpose. I will not speak
of other characteristics of the signal.

In the general case, the correspondents know nothing about
the position of each other in space. In seeking the call signals
the recipient faces a double uncertainty. He does not know the
direction of the incoming signal or its frequency. This has to be
taken into account by the sender who seeks, within the limits of
his capability, to facilitate the job of detection as much as possible. I say "within his capabilities" because the strategy of the
quest depends in large measure upon the potentialities of the
sender. Kardashev spoke of this briefly.

The task of detection becomes much simpler if the sender uses
a bandpass as wide as possible, $\Delta v \sim v$. In that case, we are
practically ruling out a frequency search at the reception end.

Lately it has become current to use compressed impulses. Compressed impulses have a broad spectrum and a high peak
power. This makes such signals very suitable for CETI call
signals. Compression can be effected either at the transmitting or
at the receiving end, in principle, but it is useless to condense it
at the sending end since propagation in the interstellar medium
will blur the signal anyway because of dispersion, with a resulting
drop in the peak flux. To compress it at the receiving end, the
recipient has to know how to do this. These difficulties may be
overcome if we use the dispersing medium between the transmitter and the receiver as a filter with a variable lag. Evidently
in that case the delay in transmission must be equal in time and
opposite in sign to the delays arising in the interstellar medium.

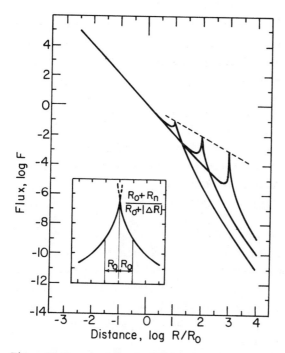

Figure 24. Focusing of an interstellar signal on a plot of the radio flux versus distance. See text for details.

In Figure 24 I show the dependence of the flux upon distance. Depending upon the delay at the transmitting end, the emission is focused at this or that point. At a small distance from the focusing point, the flux is proportional to R^{-2}, just as it would be in the absence of the medium. Then at a distance equal to two-thirds of the distance to the focusing point, compression becomes substantial and the flux increases. It reaches its maximum at the focal point; then it drops rapidly with the distance from that point; and at a much larger distance the flux declines as R^{-3}.

The dotted line connecting the peaks illustrates the fact that the flux at the focal points changes as the first power of the distance

of these points from the transmitter. The half-width of the
peak here corresponds to a halving of the flux, and is independent of the distance; it is equal to R_0, where R_0 is the distance
at which, in the absence of compensating delays, the impulse is
blurred by a factor of 2.

This system of signals makes it possible to pass from a narrow-
band to a wide-band signal for the same transmitter power and
for the same duration of the signals without any loss in the
signal-to-noise ratio. Needless to say, the impulses reach the
recipient with a large pulse ratio.

If the position of the correspondents in space is not prear-
ranged, the sender must change the focusing and send a series
of pulses that are focused at different distances. In this way, by
ruling out a frequency search at the reception end, we are com-
pelled to conduct a distance search at the transmitting end. Such
a distribution of roles in my view completely corresponds to the
purpose of the call signals, since the initiative must rest with the
sender who seeks to simplify the task of reception by his unknown correspondent.

Let us assume that we have a distant search case. Let the
sender send out a series of signals focused at R_1. He then sends
another series of impulses focused at $R_2 = R_1 + R_0$, and so on,
up to R_m. We may then ask: How in that case will the signal
look at the observation point? As the focusing front approaches
the receiver, the recipient will at some point detect a series of
pulses. These pulses indicate normal time lag; the high-frequency
component will come ahead of the low frequency. The time lag
will correspond to the distance to the focal point rather than to
the sender. If therefore the recipient seeks to determine the
distance to the source, this will turn out to be unexpectedly
small. After a time there will be a new series of impulses in
which the distance between them will be increased by one and
the same quantity, and the duration of all the impulses will de-

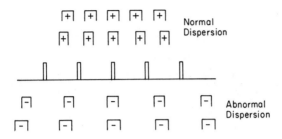

Figure 25. The time spacing of signals transmitted, as shown at center, for the cases of normal dispersion, above, and abnormal dispersion, below. See text for details.

cline, just as if the source had moved nearer to the observer by R_0. Accordingly, the peak flux will increase. A similar change of the emission parameters will be observed after successively decreasing time periods until at some point the observer detects a series of pulses with zero lag, and here the peak will reach maximum (Figure 25).

At the next moment, the observer will record the same series of pulses which will follow one another after time intervals increasing according to the same law; the distance between the pulses in each new series will continue to increase; however, the duration of the pulses, which hitherto diminished, will now begin to grow, while the peak flux will accordingly decline. But the most surprising fact of all is that these pulses will now exhibit an anomalous lag: the low-frequency components will arrive before the high-frequency ones.

These surprising properties of the emission source will probably attract the observer's attention regardless of the hypothesis concerning ETI, and prompt him to increase the sensitivity of his receiver so as to follow all the changes throughout the session—from the moment when the impulse is focused at a distance R_1 to the moment when the impulse will be focused at the limiting distance. Evidently all the physical characteristics of such a source will be studied and the distance to it determined quite accurately. If the recipient civilization is to any extent like ours, we may be certain that the theoreticians there will devise many theories to account for this phenomenon, but it seems to me that at the same time attempts will be made to decipher the information contained in these series of pulses.

Now as to the range that may be considered optimal for such a transmission. Let us assume that we have the following quantities given: first, the transmitter power P; second, the range of the distance search (R_1, R_m); third, the complete duration of

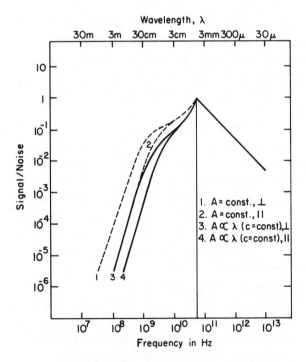

Figure 26. Signal-to-noise ratio as a function of frequency for two polarizations and two wavelength dependences.

the session including both the distance search and the direction search at the transmitting end; and, finally, the detectivity of the detection system. Since in our case we are assuming that there is no frequency search at the receiving end, the detectivity would simply be τG, where τ is the time constant of the receiver equal in the optimal case to the pulse duration and G is the antenna gain. Let us define the optimal range as the region of the spectrum where the signal-to-noise ratio reaches its maximum, provided all these quantities are in all the wavelengths identical.

In Figure 26 we see a curve for the signal-to-noise ratio as a function of frequency. The left-hand curves are due to the background noise; the right-hand, to the quantum noise. The curves intersect at the frequency of 56 gigahertz, 5.35 millimeters wavelength. This is the frequency range most suitable for the given system of call signals.

We may assume that the sender will not be too monotonous, and in each new session he will resort to a new series of pulses. In that case, the receiver will be getting new information all the time. Such signals, having fulfilled their role of attracting attention, will begin to fulfill the role of transmitting information.

Finally, to illustrate the information properties of such a system, I would like to give you this example. Let us assume that we have received five consecutive series of pulses (Figure 27). The interval of time between the pulses received, measured in the duration of a single pulse, will for these series respectively equal 7645, 4361, 4553, 4361, and 7625. If each of those numbers is expressed in a binary code and if we write one under the other and erase the zeros, we get the name of our symposium—CETI —or, as Doctor Sagan said, the constellation CETI.

PETROVICH: I would like to touch upon two matters: first, the question of communications between civilizations on pulses as short as can be. My thoughts on this subject coincide with what

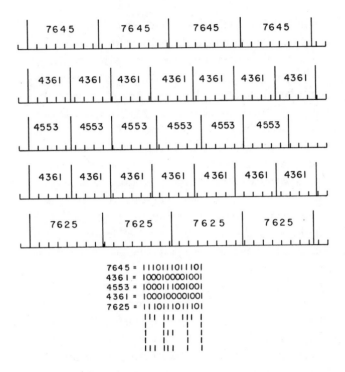

7645 = 1110111011101
4361 = 1000100001001
4553 = 1000111001001
4361 = 1000100001001
7625 = 1110111011101

Figure 27. A hypothetical interstellar message. Five consecutive series of pulses are received, each series displaying a repeated 13 digit number of binary arithmetic. Arranging these five 13-digit numbers, one beneath the other, and erasing the zeros, we find written in English the notation CETI —a surprising message from an extraterrestrial civilization.

was said here by Gindilis, but my reasons run along somewhat different lines.

Let us assume that a civilization ahead of us in development discovers a simple generation of very short and powerful pulses. This can happen in several civilizations. However, short pulses found by passing through a medium become blurred, overlap into neighboring pulses, and cannot be singled out. These civilizations evidently already know how to correct for such blurring. By pointing their transmitters at our corner of the Galaxy they can introduce frequency correction and produce the signal in a condition suitable for reception in the solar system. Therefore, it seems to me expedient, in addition to monochromatic receivers, to use receivers with much wider bands in the suspected frequency ranges.

If we organize the reception of such signals, then surely the question arises of the modulation they use. It seems to me that on the basis of what we know about modulation today, it is most probable that they use phase modulation or differential phase modulation.

In the latter case the signal can assume a periodic sequence of pulses and the information can be represented by the difference in adjacent pulse phases. The noise could be lowered by shutting off the receiver in the interval between pulses.

The second point is connected with the following: It seems to me that now it is useful to begin simulating a process of communication between two civilizations by means of computers. Let us imagine that we have two machines coupled by an equivalent interstellar medium. We can introduce interference and study the process of the establishment of communication. Computer 1 issues a simple signal or series of signals. They are distorted in the medium to the best of our ability and the second computer tries to pick out the signal from the interference. Then, we can begin teaching the computer some simple language. The computers can be widely apart, say one in Moscow and one in New York, and in this way we can begin teaching and transmitting information without feedback. Such simulation could help us to develop the theory of communication between civilizations and new languages.

Lincos may seem to many here a relatively complicated and controversial language, which could scarcely be used to teach, to instruct another party without feedback. Others think it is possible to use Lincos. The first experiments in this country to train students with Lincos have not yielded negative results. I cannot say whether the experiments are very satisfactory, but I think that it should be possible to build up a simpler language system and to convey it into interstellar space.

My last remark concerns Kardashev's proposal for the two-island experiments. I approve of the idea but it seems to me that two islands are too few. We need three islands at least, so that a choice can be made. If we have three rather than two pulses, it will give a considerable increase in compatibility. Just as in communications on the earth with three channels, the system becomes much more interference-resistant.

MORRISON: I have a short question directed at Doctors Petrovich and Gindilis. How insensitive is the depth focusing scheme to lack of knowledge of the detailed dispersive properties which vary along the line of sight?

GINDILIS: The sender produces a variable signal. You get a running wave in space and, wherever the observer is, sooner or later he will be in the focus. The number of steps will be large.

PETROVICH: I have a different answer. I take the view that focusing is conducted with respect to some fixed region of the Galaxy and there is a steady-state process. The accuracy of focusing need not be very high, since if we use relative reception of one pulse compared with another, this reduces somewhat the requirements with respect to the accuracy of focusing.

PARIISKY: There are two ways in which we can approach the observing problem. First, we can ask what sort of instrument we must use to try to make headway. Several attempts have been made to answer this question but it seems to me that in any case the answers are speculative and subjective—too many factors have to be included in the consideration. I will therefore dwell upon the second approach concerning the limiting parameters of telescopes available to man now and in the foreseeable future for observational purposes, and specifically for use in CETI.

I am a radio astronomer and so I will start with the radio astronomical part of the spectrum. Without dwelling upon the proof of this, I will postulate something about which I am sure all the radio astronomers here will agree, namely, that in this part of the spectrum the progression of techniques and new ideas in recent years justifies the conclusion that the limitations on practically all the parameters in this interval are financial. Everything depends on when we achieve this or that parameter and this, in turn, depends on when we have the money.

It would be interesting to examine the pace of radio astronomical development in terms of facilities. I will not dwell upon the details of this process and will confine myself mostly to saying that in the past ten years the capability of radio astronomy has increased by two or three orders of magnitude. We can find that extrapolation five to ten years into the future implies an increase in our potential by at least two or three orders of magnitude over such a time.

It is curious that in the optical range for a long time the limiting magnitude remained constant. There has in the past 20 years, however, also been certain progress, and we may plot a similar chart for the optical spectrum, although it is well known that here there are certain limitations. It is my personal conviction that these technical limitations will be overcome in the next 10 years. There are methods of synthesizing the image well known to radio astronomers, rapidly moving into the shorter wave region, and their extrapolation leads us to expect that similar systems will be possible in the infrared and eventually even in the optical spectrum. I would venture to remind you that this direction makes it possible to intervene into the process of image formation and rule out such seemingly insuperable obstacles as phase fluctuation in the troposphere and possibly in the interplanetary and interstellar medium.

Now a few words about specific projects in this field. It is well known that paraboloid dishes, as borne out strikingly by Von

Hoerner's discussion for the earth, and as demonstrated by Kardashev for outer space, may be very large; but they do have certain limitations. An international version of such a radio telescope with an area of 10^6 square meters (which would not, in our view, involve any insuperable technical problems) was put forward by Pulkuovo Observatory at Hamburg in 1964; and now here in the Soviet Union, while we are not building such a big telescope, we are building our first radio telescope which programs from the outset (true as its last point and not its first) the CETI problem. This is a ring-type radio telescope 600 meters in diameter intended for work in the short-wave spectrum up to 8 millimeters, with a geometric area of up to 10,000 square meters.

However, it is clear even now from what I have said that in the next five to ten years we may expect far more powerful facilities; and it seems to me that what is most promising now, at this stage at any rate, is the development of multielement systems consisting of a large number of sufficiently large elements, and at a rough estimate, with an aperture of the diameter of the earth [See *Izvestia Glavnoy Astron. Obs, USSR,* N188 (1972)].

But again, we find that we must now start thinking of larger systems. It is hard to say what sort of systems these will be. I will not speak here of such fantastic applications as the possibility of using Venus or Jupiter where the refractive index is higher and the tropospheric absorption is probably not very great, which would make it possible to look at those planets as giant lenses to view the universe, enabling us to obtain enormous collecting areas—for Jupiter, something like 10^9 square meters.

Here, I would only like to say that the use of such holographic-type or aperture-synthesis-type telescopes with reference points will make it possible to make further progress in the optical spectrum as well as in limiting star magnitudes. From the point of view of radio astronomers, optical telescopes are very poor instruments with a low efficiency; the reserves available are of the order of the ratio of the turbulence limits to the diffraction limits squared—to a certain limit, of course.

In designing CETI systems, it is necessary to allow for the very high rate of technological advance. Evidently, judging from this rate, in the next ten years the nearby extrasolar planetary systems will be available for study in considerable detail, as was the case with the solar system at the start of planetary astronomy. This trend seems to me to be a leading one, not for astrophysics alone but for the reliability of many of our CETI estimates.

And now two brief remarks: It was suggested to me by Professor Troitsky that, taking into account the dispersion effect of the medium, the focusing of the signal at some distance should be made in advance. Further it seems likely that the scale of radio telescopes will increase markedly, and that it may be possible to focus the signal at a very large distance. Although the focal length of such a system will amount to an enormous number of wavelengths, it will at the same time be many orders of magnitude less than the distance to the source.

DRAKE: It seems to me that for a 10 cm wavelength telescope whose diameter is equal to the diameter of the earth, the near field extends very far, but even though it extends only to one-tenth of a parsec, this means, I believe, that you could not focus the radiation unless you had an aperture larger than the diameter of the earth.

PARIISKY: I have in mind systems even bigger. Extrapolation shows, I am convinced, that this is not fantastic. The size of such systems will be of the order of 1 astronomical unit. The near zone of a telescope of global size can be used for three-dimensional galactic astronomy. A baseline of the order of 1 astronomical unit would give us access to the entire metagalaxy.

SAGAN: It is very interesting to hear Doctor Pariisky report that CETI problems will be incorporated from the beginning in the

600-meter telescope being built. But I wonder under what circumstances Doctors Troitsky and Pariisky would be prepared to continue for very long periods of time in such activities, if there were no evident sign of successful contact with ETI—an operational question of telescope management.

MOROZ: I will speak of the problem of choosing optimal wavelengths in the case of two-way communication with relatively near stars.

The formulas (35 and 36) were essentially presented earlier by Doctor Drake:

$$R^2 = PGA_2/4\pi P_n(\lambda),$$

and

$$G = 4\pi A_1/\lambda^2.$$

The gain is inversely proportional to the square of the wavelength. This is important. In other words, if we use antennas at the diffraction limit, the distance increases with a decrease in wavelength. I think that this is one of the fundamental limitations that we must take into account in choosing the optimum wavelength.

The second limitation is that of the dependence of the noise upon the wavelength or, rather, the noise as determined by background fluctuations. Generally speaking, if we move toward shorter wavelengths, at a sufficiently short wavelength we will have to take into account the background of the star itself. What then will be the picture for the fluctuations in background emission? This will be, of course, a very rough approximation.

Let us suppose that the frequency bandwidth is proportional to the frequency. The power of the stellar emission fluctuations increases toward short wavelengths as λ^{-2} (for wavelengths $\lambda >> 1 \mu$); the power of the general background fluctuations increases as $\lambda^{-1/2}$ for the Rayleigh-Jeans region of the spec-

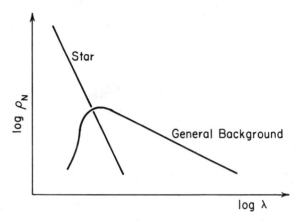

Figure 28. Schematic representations of fluctuations in the power emitted by stars and by the general galactic background.

trum, and decreases for the Wien Region. Figure 28 shows us qualitatively the dependence of these two rapid fluctuation components on λ.

Now for the signal-to-noise ratio. Here, too, we have a λ^2 factor in the denominator. Figure 29 shows the results on S/N as a function of wavelength. We get a constant where the stable background prevails while we are close to the star; and once the extended background prevails we get a drop in the S/N ratio. It seems obvious that it is best to choose a wavelength somewhere where the curve first dips, near 10^{-2} cm, because there will be no gain by moving further left and the telescope design becomes more difficult.

I have estimated the corresponding formula and I find it dependent on specific parameters: on the distance to the star, the size of the antenna, and the diameter to distance ratio in parsecs. In all cases, we then get an optimal wavelength of about 10^{-2} centimeter. This may be distasteful to radio astronomers but it seems to me justified. This leads us to the infrared part of the spectrum.

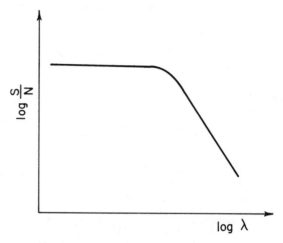

Figure 29. Dependence of the signal-to-noise ratio on wavelength for the same two factors displayed in Figure 28.

You remember, of course, what Drake showed us (Figure 18), a constant noise temperature in the centimeter range and an increase toward both longer and shorter wavelengths. I want to say that the concept of the equivalent temperature as used by us is wrong. It does not make it possible to compare the signal-to-noise ratio in short and long wave regions. In the short wave regions fluctuations are determined by

$$P_n = kT(\Delta\nu/\tau)^{1/2}. \tag{37}$$

When we have quantum fluctuations prevailing, the minimum energy is no longer kT but $h\nu$, and then we can substitute $h\nu$ for kT. This is the justification for introducing the equivalent temperature, $T_e = h\nu/k$.

But if we write the formula for the signal fluctuation, it will be altogether different. It will look like this:

$$P_n = (I_\nu\Delta\nu A_2/\tau h\nu)^{1/2}, \tag{38}$$

where I_v is the total power of the signal and A_2 the area of the receiving antenna. You see, these formulae are quite different and it is senseless to introduce this temperature to describe the noise in the visible and radio spectrum. We can show by simple examples that this procedure leads to large errors.

OLIVER: This is a misunderstanding. One can regard the quantum noise either as an increased temperature in talking about coherent receivers or in terms of the energy per quantum required to signal when talking about incoherent receivers. In either case, the high-frequency region is more costly in a way that is reflected by Doctor Drake's remarks (pp. 233–241).

MOROZ: The fact is that if we consider the specific case of nearby stars, we get a fluctuation of the stellar background much greater than the signal fluctuation. To see the signal of the background we have to make it so large that the fluctuations of the signal would be less than the fluctuations of the star. I agree with you that there are fluctuations of the signal and they have to be taken into account, but here in this narrow problem this is of little significance.

OLIVER: I propose to summarize some of the recent progress in our ideas of how we might go about the task of searching for signals that might represent electromagnetic radiations of intelligent civilizations elsewhere in the Galaxy. Most of what I shall report is the result of a summer study program conducted last year at Ames Research Center under the joint auspices of the National Aeronautics and Space Administration (represented by Ames) and the American Society for Engineering Education (represented by Stanford University), and known as Project Cyclops. So far as I am aware Cyclops was the most comprehensive study of the subject in recent years, certainly in the United States at least. This is not to say that worthwhile efforts have not been going on elsewhere, nor that these have not produced significant advances. My problem is simply that I have been too immersed in the Cyclops study to learn of other work. I therefore ask that you forgive any omissions of important advances by other workers as honest ignorance on my part. Perhaps the discussion will reveal developments of which I should have been aware, but am not.

Copies of the Cyclops report are available from NASA. The document number is CR 114445. Those seriously interested in the subject are invited to study the report, which naturally covers more aspects of the problem more thoroughly than I can do here today.

The Microwave Window

Serious study of the possibility of interstellar communication began with Cocconi and Morrison's 1959 paper in *Nature* suggesting the use of the hydrogen line at 1420 megahertz as the natural frequency on which to search for signals. This led to a very short-lived attempt in 1960 by Frank Drake at the National Radio Astronomy Obeservatory to listen for signals from two stars ϵ-Eridani and τ-Ceti, using an 85-foot antenna and a receiver with a 350°K noise temperature and a 100 hertz bandwidth. Project Ozma, as this effort was known, found no signals.

At about this same time the laser was invented and many people, beginning with Schwarz and Townes, have proposed the use of lasers for interstellar communication. Lasers have been the subject of a great deal of research during the last ten years and their ultimate capabilities and limitations are now much better understood than ten years ago. The Cyclops study made a careful comparison of lasers versus microwaves for interstellar signalling, and the verdict is heavily in favor of microwaves.

It is important to realize that microwaves are superior for *fundamental* reasons, not just because they represent a more mature art. Primarily the reasons involve energy. Any signal we use must override the natural background noises that arise from

(a) galactic noise (synchrotron radiation); (b) thermal noise (receiver and isotropic background noise); (c) quantum noise (spontaneous emission or shot noise); and (d) star noise.

Figure 30 shows the contributions of the first three of these in the microwave region. Above 1 gigahertz galactic noise falls below the isotropic background radiation. Above about 60 gigahertz quantum noise exceeds this background noise and increases indefinitely with frequency. Thus, out in space the sky is quietest from 1 to 60 gigahertz. This is the free-space microwave window. On Earth and earthlike planets water vapor and oxygen absorption lines spoil the window above 10 gigahertz, but as we shall see the low end of the window is preferable for other reasons, so this is of no consequence.

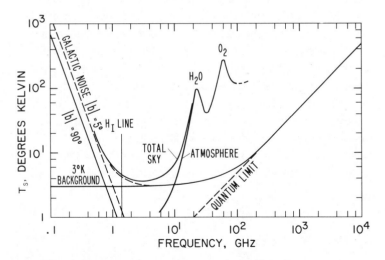

Figure 30. The system noise temperature T_s for radio astronomy as a function of frequency, allowing for quantum noise at the right. The 3°K blackbody radiation is shown toward the bottom. Galactic noise sources at two different galactic latitudes are at left.

The range limit for an interstellar communication link can be conveniently written as

$$R = \frac{d}{4} \left(\frac{P_t g_t}{\psi B}\right)^{1/2} = \frac{d}{4} \left(\frac{P_{\text{eff}}}{\psi B}\right)^{1/2}, \tag{39}$$

where R is the range, d is the receiving antenna diameter, P_t is the transmitted power, g_t is the transmitting antenna gain, $P_{\text{eff}} = P_t g_t$ is the effective radiated power, ψ is the noise power spectral density, and B is the receiver bandwidth. What can we do to maximize the range R? Obviously if we are doing the receiving we have no control over the transmitted power or the transmitting antenna gain. (Beacons will probably be radiated omnidirectionally, making $g_t = 1$.) We see that the range is directly proportional to antenna diameter, so we may need the largest antennas we can build. But antenna directivity is proportional to the diameter measured in wavelengths. If we make the directivity too great we will not be able to keep the receiver beam on the target. This says we should use the *longest* wavelength or the *lowest* frequency we can in order to have the largest usable antenna diameter.

We notice that the range is inversely proportional to the square root of the noise power per hertz (ψ), so we choose the quietest part of the spectrum. We also notice that the range is inversely proportional to the square root of the receiver bandwidth B. The narrowest bandwidth we can use is determined by the frequency drift rate of the signal. The important relations here are

$$\tau = \frac{B}{\dot{\nu}} = \frac{\text{bandwidth}}{\text{frequency drift rate}}, \tag{40}$$

$$\frac{1}{B} \approx \tau = \text{response time}, \tag{41}$$

$$\therefore B \gtrsim (\dot{\nu})^{1/2} = (\alpha\nu)^{1/2}, \tag{42}$$

$$\alpha \equiv \frac{\dot{\nu}}{\nu} = \text{frequency stability}. \tag{43}$$

Equation (40) simply says that the time a drifting signal remains in the receiver band is proportional to the bandwidth B and inversely proportional to the drift rate $\dot{\nu}$. Equation (41) says that the response time is about equal to the reciprocal of the bandwidth. If we wish the receiver to respond fully to the drifting signal, its bandwidth must be about equal to, or greater than, the square root of the drift rate. Since the frequency drift rate is proportional to the frequency itself, the minimum bandwidth we can use is proportional to the square root of frequency (Equation 42). Thus the total received noise with an ideal receiver is proportional to the sky temperature times $\nu^{1/2}$ as shown in Figure 31.

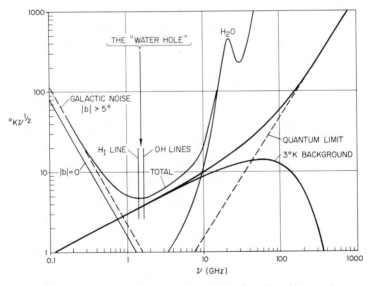

Figure 31. A plot of the total received noise of an ideal receiver as a function of frequency. The minimum noise lies between 1 and 2 GHz where the hydrogen (21 cm) and hydroxyl (18 cm) lines reside. This region, called the water hole, may be the preferred interstellar radio communications channel.

Frequency drift can originate as a result of oscillator instabilities or diurnal doppler drift. Whatever its source it will always be proportional to the operating frequency, so the new minimum shown in Figure 31 is technology-independent. We now see that the optimum part of the spectrum lies between 1 and 2 gigahertz.

To summarize, we have gravitated to the low end of the microwave window because

1. The product of sky noise and usable bandwidth is least there;
2. Larger receiving antennas can be obtained for a given minimum beamwidth than at higher frequencies; and
3. Collecting area is cheaper, the lower the frequency.

The Water Hole

The question now arises: Is there a relatively narrow frequency band in this optimum part of the spectrum where interstellar communication is especially likely? We note that the hydrogen line is right at the best part of the spectrum, so Cocconi and Morrison's original suggestion is a cogent one. However, the hydrogen line itself is noisy. Further, if we wish to transmit as well as to receive without jamming ourselves and if there are several interstellar links we might ultimately like to operate simultaneously, we would like to have not a single frequency but a naturally defined interstellar communication and search *band*.

We cannot narrow the band further than the minimum shown for *technical* reasons, so we must look for other reasons, even poetic ones. And, indeed, we find nature to be rather romantic in this instance. The hydrogen line is at 1420 megahertz. Only 242 megahertz higher in frequency is the first hydroxyl line at 1662 megahertz. Between these two lines, at the quietest part of the spectrum there are no other known spectral lines. The Cyclops team feels that this band, lying between the resonances of the two dissociation products of water is the foreordained interstellar communication band. What more poetic place could there be for water-based life to seek its kind than the age-old meeting place for all species: *the water hole?*

The Cyclops report recommends that steps be taken at once at an international level to protect this band for CETI purposes. The hydrogen line is already protected. All that is needed is to extend the protection 200 megahertz upward in frequency to the hydroxyl line.

Required Receiving Antenna Size

The best receivers we can build today add an appreciable amount of noise to that received from the sky. Instead of a sky noise temperature of about $4°K$ we will have a total system noise temperature of about $20°K$. This means we will have to collect about five times as much signal as would be needed with an *ideal* receiver. How big an antenna will we need?

We can give only a very approximate answer to this question because (a) We do not know how far into space we will have to search to find a signal, and (b) We do not know how powerful a signal to expect. Our estimates of the density of communicative life in the Galaxy are very uncertain, primarily because we do not know how long advanced races maintain an effort to communicate with one another. If, on the pessimistic side, we assume that no communication already exists and that other races like ourselves make occasional attempts at listening and sending, we may well have to search every likely star out to 1000 light years. If, on the optimistic side, interstellar communication is an existing reality, communicative races may radiate beacons to attract new races to the Galactic community and they may do this for millions of years. In this case we may detect other life closer than 100 light years. Or we may detect leakage signals within this range.

As to the power radiated by beacons, we might assume that races typically balance the cost of transmission against the cost

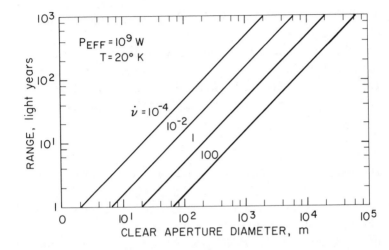

Figure 32. The range capability as a function of aperture diameter for frequency drift rates and other parameters as shown.

of reception. In this case beacon powers on the order of 10^9 watts are not unreasonable. Figure 32 shows the range capability of a receiving system plotted against antenna diameter. The system noise temperature is assumed to be $20°K$ and the bandwidths are assumed to be matched to the frequency drift rates shown.

Uncompensated doppler drift rates are expected to be on the order of 1 hertz per second (at the water hole) and compensated rates might be as low as 10^{-2} hertz per second. We see that to detect a 1000 megawatt beacon out to 1000 light years we will need antenna diameters of several kilometers. Single antennas this large cannot be constructed on Earth and are prohibitively expensive to build and operate in space.

We are therefore forced to consider phased arrays of smaller antennas—perhaps 1000 to 10,000 dishes 100 meters in diameter. An array offers two distinct advantages over a single antenna. First, the precise pointing of the beam is done electrically rather than mechanically. Pointing precision of less than one second of arc is difficult mechanically but easy to do electrically. Second, the array can be gradually increased in size with time. It need never be built bigger than needed to do the job. There is no danger of overdesign. In view of our uncertainty as to the size required this is an important advantage.

Limitations on Array Size

In a phased array the signals received by each individual antenna in the array—by each *element*—are heterodyned down to a common intermediate frequency for transmission to a central headquarters where the signals are added together. To operate satisfactorily—

1. The local oscillators at each receiver must have the correct relative phase relationship within a few electrical degrees.
2. The intermediate frequency signals must all be transmitted to the central headquarters over circuits having constant gain and delay.
3. Controlled variable delays must be introduced into all these paths to steer the beam. These delays must be accurate to a fraction of a nanosecond.
4. All receivers must be remotely tunable to the same band and function without attention over long periods.
5. The array must be self-checking and self-calibrating under computer control.
6. The whole search sequence must be automated.

A large part of the effort of the Cyclops study was devoted to solving these problems. As a result of the study I can now report with confidence that arrays up to 10 kilometers in diameter are practical up to frequencies of 10 gigahertz and arrays up to 30 kilometers in diameter are practical up to 3 gigahertz. Because the antenna elements must be spaced about three times their diameter these correspond to clear apertures 3 kilometers and 10 kilometers in diameter at 10 gigahertz and 3 gigahertz, respectively. We also believe that remotely tunable receivers with noise temperatures as low as 20°K are feasible at 1 to 2 gigahertz and 30°K appears possible at 10 gigahertz. The Cyclops receiver design permits an instantaneous bandwidth of 100 megahertz. With further study this could probably be increased to 200 megahertz.

I do not have time to go into the details of how all this is done. Those interested are invited to read the Cyclops report. For the present I must merely assert that I now believe that this kind of receiving system is possible. Future advances in technology may make the job easier but it can be done today.

Thus we now visualize the search system as composed of 1000 or more large steerable radio telescopes all connected together to form one huge antenna; an orchard of antennas covering 20 square kilometers or more, all feeding a single data processing system and all under automatic computer control. The system could begin searching nearby stars with only a few antennas operating and then carry the search farther and farther into space as more antennas are added to the array. The total construction time could be 10 to 20 years.

The Search Strategy

The very high directivity of the array—in fact, of each element of the array—precludes the possibility of searching blindly in space over the entire sky, or of searching several stars at one time. Instead we must compile a list of likely target stars and search these one at a time.

The likely stars are believed to be late F, all G, and early K main sequence stars. Within 1000 light years of the sun there are about one million such stars. No catalogue of these stars exists. Hence before a really deep space search can begin these stars must be identified and located by an optical search program. The Cyclops report offers some suggestions as to how this may be done.

Lists do exist for the nearest few thousand stars. These could be searched while the complete list is being developed. For the nearer 1000 or so likely stars, i.e., for all target stars out to about 100 light years, we can afford to devote several hours or days per star. This means we can search not only for beacons but for leakage signals as well—the signals "they" radiate for their own purposes.

But if we are going to search the million or more stars out to 1000 light years in a reasonable time—say 30 years—we can only devote about 1000 seconds to each star. Thus we are forced to depend on beacons that are always shining—steady signals that are always present.

Combing the Spectrum

If signals are to be detected whose drift rates are 0.01 hertz per second to 1 hertz per second the corresponding receiver bandwidths are 0.1 hertz to 1 hertz. Let us assume the lower drift rate and a receiver bandwidth of 0.1 hertz. Then the response time for the receiver is 10 seconds and to search 100 megahertz of spectrum would require that we spend 10 seconds at least on each of 10^9 channels. This would mean a search time of 10^{10} seconds or 300 years per star. One of the major achievements of the Cyclops study was the reduction of this search time per star from 300 years to 1000 seconds.

The secret of this fantastic time reduction, or search speed increase, is to make use of the enormously high-speed Fourier transforming capability of an optical system. I would like to spend a few minutes now to show you how this system works.

Most engineers today are familiar with the fact that the electric field amplitude distribution in the image plane of a lens is the

Fourier transform of the complex two-dimensional amplitude distribution over the aperture plane of the lens. This fact is exploited in holography and image filtering and in aperture synthesis in radio astronomy. Not so well known is the fact that this full transforming power can be used on a one-dimensional signal such as the intermediate frequency signal from an antenna array. This possibility was pointed out by Robert Markevich of the Ampex Corporation.

The signal to be analyzed is first separated by a series of mixers and filters into several baseband signals each about 1 megahertz in bandwidth. Each of these 1 megahertz wide signals is biased so as to avoid negative values and then recorded in raster form on a strip of film as shown in Figure 33. Here we show the signal applied to an optical modulator so as to modulate the intensity

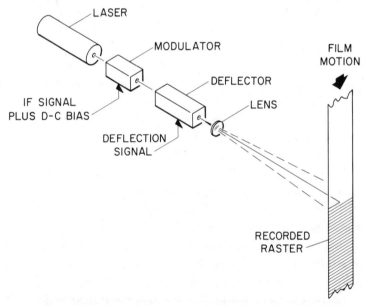

Figure 33. Initial stage of a high-speed Fourier transforming optical system for rapid spectrum searches for interstellar signals. See text for details.

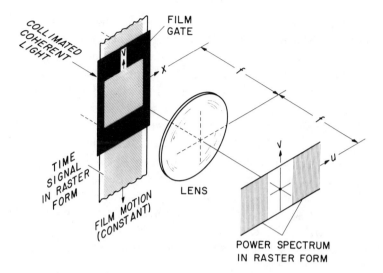

Figure 34. Second film processing stage for high-speed frequency searches. See text for further details.

of a laser beam, which is then deflected in a sawtooth manner across a moving strip of photographic film. The film is then processed in a rapid processor and passed through the film gate of an optical spectrum analyzer where the whole window is illuminated by coherent light from another laser as shown in Figure 34. The film gate is at the front focal plane of a lens. The light intensity in the rear focal plane then consists of the power spectrum of the recorded signal. Here nature is very kind to us for this power spectrum is also displayed conveniently in raster form, and not once but twice!

To understand how this happens let us look at Figure 35. Let us imagine that the IF signal consists of a single coherent sine wave and let us imagine that there is exactly an integer number n of cycles per scanning line. Then the dark and light segments of each line will lie directly under those of the line above and the film in the analyzer gate will consist of n dark and n light vertical bars as shown in the upper left-hand figure. This bar pattern with the vertical axis a-a' acts as a diffraction grating, so in the transform plane we will find three bright spots: a spot at the origin O, due to the dc component added to the IF signal, and a spot at A_1 and A_2 on either side, due to the sinusoid.

Suppose now that we increase the frequency of the sinusoid slightly. Now the dark and light segments of each line will lie slightly to the left of those in the line above. We still have a diffraction grating but the axis a-a' is now tilted through an angle θ. The spots A_1 and A_2 also rotate through the angle θ, A_1 moving up and A_2 moving down. Their horizontal separation is increased slightly because the horizontal pitch of the grating is now slightly less. Notice also that a new grating with axis b-b' is beginning to appear.

If the frequency of the sinusoid is now increased until there are $n + \frac{1}{2}$ cycles per line, the dark and light segments of each line alternate from line to line like bricks in a brick wall. Both grating axes are now equally prominent. In the transform plane, as the spots A_1 and A_2 (due to axis a-a') pass out of the field, two new spots B_1 and B_2 (due to axis b-b') appear to take their place.

As the frequency is further increased the axis b-b' becomes more and more nearly vertical and the spots B_1 and B_2 approach the horizontal axis in the transform plane. When there is again an integer number of cycles per scanning line, the spots B_1 and B_2 arrive at the horizontal axis but at points a little farther from O than the original spots because there are now $n + 1$ cycles per scanning line and the grating has a finer pitch.

As we continue to increase the frequency, pair after pair of spots trace out two rasters in the transform plane. The series of patterns on the recorded film are like the bar patterns produced by a sine wave oscillator on a television set and the spots in the

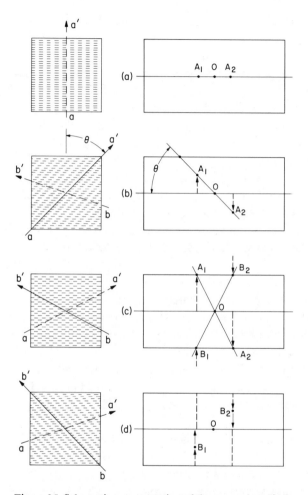

Figure 35. Schematic representation of the convenient display of the power spectrum (see Figures 33 and 34) in raster form. See text for details.

transform plane measure at all times the x and y pitches of this bar pattern.

Since the process is linear, superposition applies. The entire power spectrum of a complex signal will be portrayed at all times in the transform plane.

The frequency interval per line in the power spectrum raster is the scanning frequency used in the recording. The resolution of the power spectrum is the reciprocal of the time represented by the signal in the gate.

If a 1 megahertz bandwidth signal is recorded with a scanning frequency of 1 kilohertz and there are 1000 lines or 1 second of recorded signal in the gate, the resolution of the spectrum will be 1 hertz. If the scanning frequency is 316 hertz and 3160 lines or 10 seconds of signal are in the gate the resolution will be 0.1 hertz.

This is about the present state of the art. A single optical analyzer will resolve 1 megahertz of signal into 10^7 channels each 0.1 hertz wide. Thus to resolve the 200 megahertz of the "water hole" into 0.1 hertz channels we need 200 analyzers for each received polarization.

The film usage is moderate. It depends only on the *total* bandwidth to be analyzed and the resolving power of the film, and is about 20 square centimeters per second for a 200 megahertz band.

Suppose now that there is only noise in the IF signal. Then there will be only noise in the power spectrum. But if a coherent signal is also present there will be a persistently bright dot on one of the raster lines of each power spectrum. If we were to photograph *that scanning line only* as frame after frame of film passed through the analyzer, and if we were to arrange these successive photographs in a raster we might get a picture such as that shown in Figure 36. Here we see a coherent signal that is drifting in frequency. We notice that this signal is not particularly evident

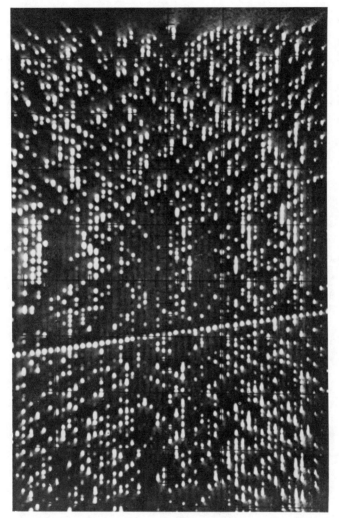

Figure 36. Final representation of a frequency drifting signal detected by the high-speed Fourier transform method discussed in the text.

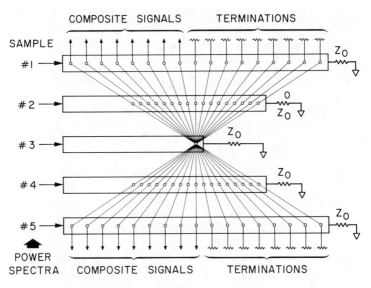

Figure 37. Delay line analysis for all possible drift rates to detect coherent signals. See text for details.

on any single raster line of this picture but is clearly evident in the total picture.

If we were to scan this picture with a slit oriented exactly parallel to the line in the picture, we would get a large pulse as we crossed the line. This is the principle used in the Cyclops data processing system.

The power spectra from all the optical analyzers are imaged onto vidicon camera tubes, and then scanned along the power spectrum raster lines. The video signals are recorded on magnetic disks. Each frame is recorded in real time, say 10 seconds for 0.1 hertz resolution. After 100 frames or 1000 seconds of power spectra are recorded, all the frames are played back simultaneously.

The video signals are sent down delay lines as indicated in Figure 37. Rows of taps are disposed across the hundred delay lines (only 5 lines are shown) and the signals picked up by each row of taps are added. The rows of taps are slanted at all possible angles to detect signals having all possible drift rates. On one set of taps—the one for which the slant matches the drift rate—all the recorded pulses from the coherent signal will add in phase and cause that detector to exceed the threshold possible from noise alone. When this happens a coherent signal was probably present.

Figure 38 shows the performance of the Cyclops detector. The curves are drawn for an overall false alarm probability of 10 percent per 100 megahertz of bandwidth. The numbers on the curves are the probabilities of missing the signal. We see that with 100 integrations the probability of missing the signal is only 10 percent if the ratio of received signal power to noise power in the resolution bandwidth of the analyzer is unity, or 0 decibel.

This means that with 1000 seconds of observing time per star we can reliably detect a signal whose power is equal to the noise power in a 0.1 hertz band, or 90 decibels below the power in the

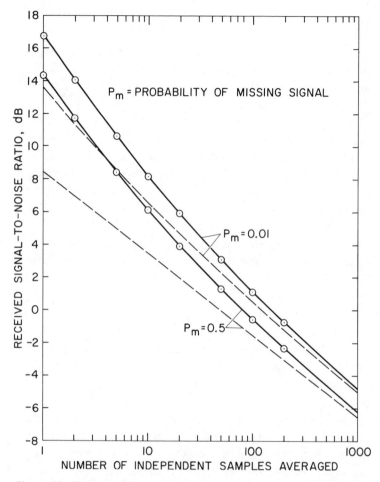

Figure 38. The probability of overlooking a signal with the proposed Cyclops detection system, as a function of the signal-to-noise ratio and the number of independent samples averaged.

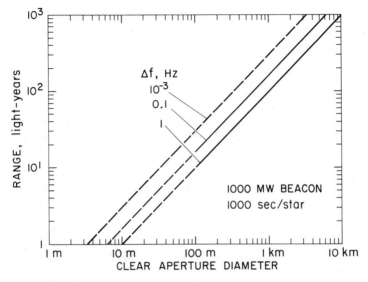

Figure 39. Overall performance of the proposed Cyclops system.

Table 5. Comparison of Ozma with Proposed Cyclops System

	Ozma (1960)	5-km Cyclops (1971)
Antenna Diameter	27 m	5000 m
Antenna Efficiency	0.5	0.8
Noise Temperature	350°K	20°K
Bandwidth	100 Hz	200 MHz
Channel Width	100 Hz	0.1 Hz
Integration Time	100 sec	10 sec
Spectral Search Speed	1 Hz/sec	2×10^5 Hz/sec
Limiting Sensitivity	1.7×10^{-23} W/m^2	1.7×10^{-30} W/m^2
Figure of Merit	1	2×10^{12}

full 100 megahertz received band. We know of no other detection process that comes within orders of magnitude of matching this performance.

Overall Performance

Figure 39 shows the overall performance of a system with a 20°K noise temperature using the proposed Cyclops detector and constant search time of 1000 seconds per star. The lowest curve assumes a resolving power of 1 hertz in the spectrum analyzer and therefore that 1000 spectra are added. This would allow doppler drift rates of up to 1 hertz per second. The middle curve assumes a resolution of 0.1 hertz and the addition of 100 spectra. This would allow drift rates of up to 0.01 hertz per second. This is thought to be maximum limiting performance. The upper curve assumes a channel width of 10^{-3} hertz and no addition of spectra. Here the doppler rate would have to be 10^{-6} hertz per second or less.

We see that to detect a 1000 megawatt omnidirectional beacon at a range of 1000 light years we would need an antenna having about 5 kilometer clear aperture.

Table 5 compares the performance obtainable with the Ozma system used by Frank Drake in 1960 with a proposed 5 kilometer Cyclops system. We see that the Cyclops system can search the spectrum 200,000 times faster and has ten million times the sensitivity. The product of these two factors gives the figure of merit of Cyclops over Ozma of 2×10^{12}: It is of some interest to note that the limiting sensitivity of the Cyclops system is about two photons per second per million square meters. No optical system could come close to this sensitivity and optical photons are 100,000 times more expensive (i.e., more energetic).

Not all of the increased performance of Cyclops over Ozma is due to technological progress. We must remember that the Ozma system cost on the order of 1 million dollars while the 5 km Cyclops system would cost over 10,000 million dollars.

Shall We Try?

It appears from our calculations that present-day technology is capable of mounting a very effective search for extraterrestrial signals. The question we must ask ourselves is: Is it worth about half the cost of the Apollo program to attempt such a search?

The biggest barrier to making such an expenditure is that we cannot *guarantee* success. Such an undertaking represents a very expensive gamble. Against the risk we must weigh the potential benefits. In all likelihood intelligent civilizations have existed in the Galaxy for four or five billion years. It seems very probable that many of these have established interstellar contact. If so, then communication has been taking place between civilizations for aeons, and beacons may well exist to help young races, such as ourselves, to join this Galactic community.

Some of us who have studied the problem for a long time feel that all past human history may indeed be merely a prelude to an inconceivably exciting future as participants in a Galactic culture. At the very least we could expect to gain access to a heritage of knowledge aeons old. What astronomer would not cherish photographs of the Galaxy and the universe taken 5 billion years ago? Wouldn't we all like to learn the natural histories of all life in the Galaxy and the social structures that have led to the survival of the oldest cultures?

I cannot conceive that man, unable to travel to the stars, will not some day attempt to reach other life through interstellar communication. Is it too early to begin the search? Or will the year 2001 find us no longer an isolated species but beginning a new epoch in the evolution of life on earth?

BRAUDE: I would like to ask Doctor Oliver: Why so many antennas when you can use a system where you have two mirrors with broad beams. By arranging them with cables, you can use them to receive signals from very large areas of the sky. It is possible, given several sets of receivers, to obtain practically any number of beams looking in different directions at different parts of the sky.

OLIVER: The large number of receiving antennas which I quoted are simply required with present-day structures to get the required total aperture. The techniques of aperture synthesis which can yield high-resolution pictures in radio astronomy are simply not applicable to the interstellar communications problem where one needs simultaneous large collecting apertures. We did study, in the Cyclops Project, systems for imaging as much of the radio sky as was possible because of the limited field of view of each element. Within a thousand light years, there are far less than one star per field of view of the elements of the array. Thus, these imaging schemes do not seem to offer much for simultaneous search of many stars unless one goes to substantially greater ranges. But they do offer great advantage for pointing the array properly and for sky survey.

BURKE: I just wanted to give a brief example of the role of technology in the radio range. The Westerbrook antenna at Leiden in the Netherlands is an array following much of the philosophy that Doctor Oliver and others spoke of. It does a Fourier analysis of the sky. In Figure 40 is a radio image of M51, the well-known spiral galaxy with a companion galaxy close by. The picture was made by W. Jaffe from the data of Mathewson et al.*

MINSKY: What frequency?

BURKE: By processing the Fourier components and then redisplaying the intensity contours on an oscilloscope, a picture of the sky at radio wavelengths was developed at the observatory at

* D. S. Mathewson, P. C. van der Kruit, and W. N. Brouw, A High-Resolution Radio Continuum Survey of M51 and NGC 5195 at 1415 MHz, *Astronomy and Astrophysics* 17 (1972): 468–486.

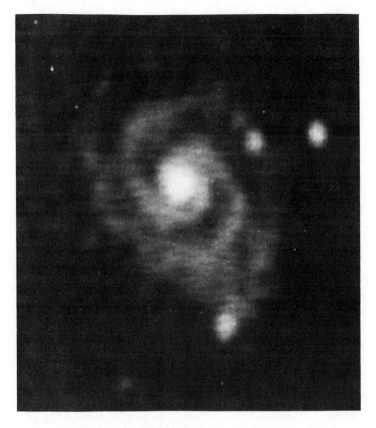

Figure 40. Aperture synthesis Fourier reconstruction of the spiral galaxy
M51, from the data of Mathewson et al.

a wavelength of 21 cm. This is the first such radio picture.

TOWNES: That is a line spectrum?

BURKE: No, this is a continuum spectrum, continuum nonthermal synchrotron radiation from M51. It is resolution-limited at the present time.

TOWNES: It is very pretty. What is the resolution?

BURKE: The resolution is 22 seconds of arc.

The only additional comment I would like to make is that the original choice of frequency was made only a few years ago on very rational technical grounds such as people have been discussing here. They are now constructing front ends for three times the present frequency, because it is quite clear that the continuum results will be much more interesting there. I mention this as a caution against assuming that the first and second laws of thermodynamics are a guide for an exact choice of wavelengths.

GOLD: A general comment on questions of transmission and reception: Doctor Oliver and others mentioned that we should attempt to do both. Mere reception would give information now. But as soon as I find something of interest, transmission will give me information a few hundred years from now. I am reminded of a quotation from the House of Commons some years ago: "Why should I do this for posterity? What has posterity done for me?"

TOWNES: It seems to me some of the crucial points here are, first, that communication is now possible. The second point is that we really know nothing much about the probabilities; the number of life-bearing systems in our Galaxy might be anywhere from none to perhaps 10^8 or 10^9. If one believes in Bishop Berkeley's philosophy, of course, this civilization is nonexistent. The third particular point that I think is crucial is the very rapid increase in technology, at least within our own civilization, and perhaps in other forms of intelligent societies. Everyone knows this almost vertical rise of technology with time, and yet I think we have not yet taken adequate account of it in our discussions. Next is that we see now, at least, that there are a number of techniques which can be successful in communication and I believe we may suppose there are others that we haven't yet imagined.

Finally, I would mention the large number of areas where subjective judgment comes into play. It almost makes the field a nonscience, but nevertheless one still exceedingly interesting. For example, one of the subjective judgments came into play in the discussion between Doctor Sagan and myself (pp. 210–211), Sagan believing that civilizations much advanced beyond us would in general not be interested in communicating. He may be right, but my own judgment is that our experience indicates the reverse. In particular, we are especially interested in communicating with Stone Age peoples, which may be a thousand years behind us in technology. These are not ants but rather people, individuals who have almost the intellectual capability we have. We are so interested that we frequently destroy such civilizations. Nevertheless, I recognize I may be wrong. I simply note that there are these insoluble differences in judgment.

In such a situation, it seems wise to me to make a number of attempts in a number of different directions, because we cannot be sure of any. This is a primary precept with me: Essentially, why do a hard experiment when there are a number of easy ones that need doing? So, I would think one ought to look at all reasonable frequencies where the looking is not too expensive and where the number of different techniques allow for the individual judgments of individual scientists as to what they should do.

Furthermore, the great value of this approach is that our problem bears on so many different and important sciences. Someone might, for example, try to emphasize the detection and study of planets around other stars. Someone else might emphasize the evolutionary history of planets and of stars. Another might emphasize the evolution of life on this planet, that is, from a basic biochemical point of view. Others might emphasize the efficient study of radiation and its handling, or perhaps spacecraft technology. It seems to me that all of these may contribute very importantly and in any one of them we may get an exceedingly useful answer which will clarify things; in themselves they are a very valuable contribution to science which we can well afford to support.

Let me speak a little bit specifically about the laser field. I was very much impressed by Doctor Oliver's report, which represents a very valuable and useful study, particularly of the radio region. I don't really feel that the right kind of effort and imagination has yet explored the possibilities in other frequency regions. Furthermore, the rate of development is a very important factor. Professor Pariisky has stated that one gains two or three orders of magnitude in (essentially) signal-to-noise during every five or ten years. Perhaps the slowest development, but yet important, is the total rate of productivity of the human race, which doubles perhaps every ten years. These things in themselves mean, it seems to me, that one should emphasize the closer stars, because to communicate with stars let's say 1000 light years away or more means that for the round trip we must be using technology that is very outdated. Such communication could easily wait 100 years and with our rate of growth might therefore be much easier; perhaps we will eventually reach some kind of saturation in the exploration of the physical possibilities.

Perhaps one of the most rapidly advancing fields of all is that of quantum electronics and lasers. During the last decade the transmitter power that is available from lasers has increased an order of magnitude almost every year. I see no reason why this should stop, so we might say at least an order of magnitude

every two years for some time. I noted that Doctor Drake took
a figure on his chart for lasers which was outdated by, I think,
about two orders of magnitude already. The present maximum
continuous wave power that had been published I believe is 30
kilowatts, and I think Doctor Oliver's proposal of calculating
with 1 megawatt is a very reasonable one for the almost immediate future. Basically, I see no special upper limit for this
power, and here one gets into the question of what quantities
are optimized. Most of the discussion today has tried to optimize
the power transmission—that is, minimize the amount of power
required. But if one recognizes that there is some cost in materials and construction, then perhaps there may be other things
which need to be optimized. The amount of power may be indefinitely large and relatively cheap. Another kind of consideration in this same direction is, for example, wind problems. In an
atmosphere where there is a very intense wind, one may want
perhaps to minimize the size of an antenna and to maximize its
rigidity.

A second area where there have been some recent advances
for lasers is in the coherence of transmission through the atmosphere. We have recently made some measurements which show
that a 120-inch telescope, for example, is diffraction limited at
5 microns; in other words, the aperture which is inherently
usable increases considerably more rapidly than linearly with
the wavelength. So, I would be tempted to use antennas at 10
microns—considerably longer wavelengths than Doctor Oliver
is assuming. A 10-meter antenna for 10 microns seems to me
very reasonable and not very expensive, and probably would be
diffraction limited, although we don't know the maximum size.

Doctor Pariisky pointed out that new techniques allow further
elimination of the trouble of phase incoherence. In about two
years there will be a 10-micron transmission from a satellite
which will allow further study of the atmospheric properties

and which, in particular, will allow, if one wishes, a correction of the phase difference between parts of an array of antennas, because this gives a perfect test signal for correction. So, there are many possible improvements in this direction, too.

I mention these and the fast rate of change of technology primarily because I think it hazardous to try to make absolute decisions, or decisions with much absoluteness at this point. Within five years our own possibilities will change and any civilization that may be one hundred years beyond us may see things really quite differently. All of these times are very short compared with the time scale we must consider.

I can imagine an advanced civilization monitoring very carefully at all frequency ranges every star and planet anywhere near it, possibly for defense purposes, to give warning, possibly simply from curiosity to make a thorough study of every possible change in nearby stellar and planetary systems. Again, let me emphasize what I consider the priority of the closer stars. Not only are these easier for us to examine but presumably we will be of more interest to those on the nearby stars.

All of these comments are really matters of emphasis. I don't take issue with the specific science that has been stated except that I would take rather different numbers in some cases. But there is one particular point where I think some of the science has not been set in the right perspective at all. That is with respect to noise in the submillimeter infrared regions. The basic noise which Doctor Oliver discussed is certainly quite correct for any kind of linear detector; but for a nonlinear detector, such as a photon detector, he only indicated in words that there is a substantial difference. While a simple statement may lead to just the things which Drake and Oliver have put down, I think there is an important additional point. If one were to calculate signal-to-noise for one photon, then the noise must come in as

$\exp(-h\nu/kT)$, which, for a 3 °K temperature and a 10 micron photon, say, means about 10^{-200}. This again is too simple a statement, however. What it says is that if one detects a single photon, the probability of that being a mistake is 10^{-200}.

The advantage in this is that with this idealized system, one can search many modes simultaneously. One could search 10^{200} modes and have fair probability of a signal-to-noise of 1:1. Thus, in designing a new system one could search the whole 10-micron spectrum all at once with no background noise; so you see these considerations make quite a difference.

On the other hand, for transmitting information, then the impression one gets from the formulae is essentially right. That is, for searching this has a substantial advantage for transmission of the information and one must take a noise temperature of $h\nu/k$.

It is still more complicated because this factor is not yet correct. There is, of course, solar radiation, stellar background radiation, and so on. So I think from a practical point of view these background radiations will give instead of perhaps 10^{-200} noise, maybe 10^{-14}; this seems to me an important factor in affecting the speed of possible search.

On the relative advantages between laser and radio frequencies, I think Oliver said a number of essentially right and pertinent things. But there is this different point of view, that if we want to communicate with, let us say, stars out to five or ten thousand light years only, then lasers as well as other methods are really quite practical. I think the millimeter and submillimeter range, as pointed out earlier, is also a very reasonable possibility. Again it depends on what one optimizes. But I would maintain that a 10-meter antenna for 10 microns is probably more advantageous than a 1-kilometer antenna for microwaves. This is a tradeoff only under certain specific assumptions. Under different assumptions, there would be a different tradeoff.

I believe a very substantial advantage of microwaves, which wasn't perhaps mentioned forcefully enough, is that if one wishes to look over a very large number of stars at a long distance, almost any microwave beam will encompass a large number of stars at one time. I think there is a substantial advantage to including in the beam a large number of stars if one wishes to look a very long distance. For closer distances, such as a few thousand light years, I don't think that is so important.

I emphasize lasers this much not because I would like to sell lasers but, rather, as an illustration of other possibilities. It is important to recognize that we may not have thought imaginatively enough or with enough conviction about some of the other methods.

I would also like to mention one further kind of technique which I don't understand very well, but I don't understand why it isn't a reasonable one, and that is the actual sending of small spaceships. I really wonder whether that has been explored sufficiently well and whether a distant civilization, if it seriously wanted to contact us, might not send a spaceship with appropriate flares and pictures, and so on. I am not familiar enough with the numbers to be prepared to argue that this is completely reasonable, but I see no reason why it is not at the moment. So I hope we will all look in a variety of ways, and discovery may come tomorrow. It may come in some very unexpected way through a completely different route.

DRAKE: I would like to return briefly to the question of the comparison between optical and radio communication systems. Professor Townes has called to our attention the rapid growth in laser power levels. I believe that this rapid growth to some extent, and perhaps entirely, reflects the fact that the laser was an invention that occurred late, long after the world's technology was ready for it. It is now rapidly overcoming that late development time to catch up with the other methods of power generation, and so this rapid growth will not be sustained but, rather, as the laser approaches in power the radio generation methods, the rate of growth will level off to that which is experienced by radio transmission.

Indeed, after the power levels have reached some high level such as radio has achieved, one reaches a situation where ever greater power levels are most economically achieved by joining systems together in electrical "parallel," if you will. In this way, the cost of higher power systems goes linearly with the power level, which is a very good law.

What this says to me is that one can make a fair comparison between laser systems and radio. If you take into account that lasers may soon reach 1 megawatt, as Doctor Oliver and I suggested, you may compare the effectiveness of systems of similar cost using such transmitters in the two wavelength regions. In fact, the example I gave earlier (pp. 234–238) was based on this concept. I mentioned what happens if you use a 1-megawatt laser instead of a 1-watt laser.

To give you an example again, the Arecibo telescope with a 1-megawatt transmitter costs about the same as a 200-inch telescope. One can be generous toward the laser systems if one says that a 200-inch telescope plus a 1-megawatt laser is equal in price to an Arecibo reflector with a 1-megawatt transmitter, and that larger systems in each case could be made by "paralleling up" such systems. Of course, the relative advantage or disadvantage would remain unchanged.

Such a system as I described goes in favor of the radio technique—the Arecibo telescope plus the 1-megawatt transmitter —by a factor of 10^3. I believe that even much larger systems, for the reason I have given you, would retain about the same relative advantage for a given monetary expenditure.

Of course, there can be technical developments which shift the balance either this way or that, but a factor of 1000 is very hard to overcome. So I would say it is probable, although not certain, that the advantage is in favor of the radio system.

TOWNES: I think certainly the right kind of answer on power is to recognize that we have in the immediate future a laser power limited primarily by the cost of copper. There are other factors, though, which I think have not been properly taken into account. One, for example, is this factor of as much as 10^8, possibly 10^{10}, in the effectiveness with which one can search, which really is not in the formula as given by Doctor Drake; a factor of 10^8 or 10^{10} is not something to change things substantially. The search mode is not the only mode and if one talks about information, then the formula as Doctor Drake gave it is, I believe, quite correct. Yet initially the search mode is perhaps the most important one. When something is located and can be found, 10^8 is a rather large factor.

On costs, I would also point out that it is exceedingly misleading to use the cost of the 200-inch telescope as a guide to the cost of a 10-micron modern telescope. I would propose, for example, to stamp out such telescopes. We have, in fact, made mirrors this way in the laboratory which were quite good. The initial cost of making one surface might be comparable with that 200-inch telescope, although techniques have evolved considerably now and a precision for 10-micron wavelengths would of course be much less than anything we have for optical wavelengths 20 times shorter.

One then gets into the question of cost of material. If stamping out things is possible, it is the cost of material which might be limiting, or I would say fragility of structures in storms which might be limiting. If cost of material is the limiting factor, then the size of the antenna becomes very important and that strongly favors short wavelengths. Again, I don't pretend to give the answers completely; I simply point out that there are many factors here which do not yet, I think, balance as one might possibly balance them.

OLIVER: I would like to ask Doctor Townes to comment on two points. I agree that there can be clever methods of making mirrors such as you suggest, but we have to realize that in the area of optical frequencies we have to achieve essentially the same receiving area at whatever frequency we are talking about, so I don't see a great cost advantage there, though there may be one.

The second point that bothers me is this: I think you concluded that we could search a very wide spectrum at a time on the basis of the very rapid falloff in the background caused by the cutoff of thermal radiation, but to me this neglects the enormous background of the primary star of the planet we are attempting to detect. Do you have a way of avoiding this background?

TOWNES: Let me answer the second question first. The $3\,^\circ$K blackbody radiation would give one an advantage on this exponential formulation of 10^{200}. That is unrealistic because of the background of the star. If one allows for the background of the star, the advantage is 10^8 or 10^{10}. If one is uncertain where the planet is, then looking in the dark regions of space at the site of the star for such a low background costs nothing.

On the first question as to size, we know there are two factors: For the receiver, the total area certainly counts. For the transmitter it is a comparison of area and wavelengths. One must take into account both factors in judging what the overall system would be. Even though we might expend only the cost of the receiver, someone else has to make a choice on the transmitter. They might, for example, assume that we are fairly advanced and try to use the cheapest and smallest system on the transmitter end. I think both of these have to come in. How one weighs them is a matter of choice as to whether we put the burden on them or the burden on ourselves.

MINSKY: I assume that Doctor Townes is worried about storms or such on their planet.

TOWNES: Exactly.

MINSKY: On Jupiter or someplace like that, you might not be able to build an antenna bigger than a certain size.

OLIVER: With such storms, I doubt if the atmosphere would be conducive to life.

SAGAN: I would conclude just the opposite.

KARDASHEV: In the infrared and submillimeter region, we cannot observe such a low background value. This region is poorly studied but we are now quite familiar with the fact that in this region there is thermal emission of interstellar dust and many molecular lines.

GINDILIS: We are not taking into account an important factor—interstellar absorption. When we observe in the Galactic plane, at optical wavelengths, the absorption is equal to two stellar magnitudes per kiloparsec; that means that at a distance of 2.5 kiloparsecs the intensity is attenuated one hundredfold. Consequently, if we want to observe the whole Galaxy, we must rule out, owing to opaqueness, wavelengths up to 50 or 60 microns.

MOROZ: If you look at the formula (equation 39) written by Oliver, you will see that it includes the g factor, which in the general case is proportional to the square of the frequency. If we accept both formulae without discussion, the first is undoubtedly correct but as to the second, I have some doubts.

Even if it is impossible to operate in the submillimeter range from the earth, in thirty or forty years this will no longer be the case. I think the tens to hundreds of micron range is a very promising part of the spectrum for CETI work. We can get much more directionality there.

As to the background temperature mentioned by Kardashev, I think the prevailing feeling is that in some directions the background can be of the order of $100\,^\circ$K. There are dust clouds, but in other directions possibly not; there may be directions where the background may be only tens of degrees.

Incidentally, in the center of the Galaxy there is an extended source of maximum emission in the 100-micron region. Could this be a dialogue between two ETI's?

MESSAGE CONTENTS

PANOVKIN: The central problem in CETI, as I see it, is that of the possibility of the establishment of contact with extraterrestrial intelligence. There are three conditions for this problem. First the energy aspects of the matter. This has been discussed here at some length. The second expressed condition is a necessary structural variety necessary for the transmission of information; that is, there must be a link with structural information. I will call this the purely structural aspect of the problem. This, too, has been dealt with in several discussions about the optimal coding and recoding of signals, redundant information, and[1] similar problems.

There is also, however, a third expressed condition for CETI. This is the possibility of transmitting substantive semantic information—the possibility of understanding your correspondent, understanding what he is driving at. As far as I can see, within the framework of CETI this third problem has not received proper attention. But it is, I repeat, one of the three expressed conditions for communication.

There is a certain subject matter which is to be discussed by two correspondents—the objective contents of the communication. Now, what are the operations that have to be performed by each of the correspondents to establish meaningful communication? First of all, the transmitting correspondent must reflect in his mind the object he wishes to communicate about. Once reflected in his mind, this object has to be coded in order to be transmitted. It has to be coded in a system of symbols, and that requires a second operation of equal importance, an operation which I will call designation—that is, the real contents have to be translated into a set of symbols, coding. Now everything is ready for the actual transmission. Semantically speaking, this operation, too, has two phases, the energy and the set of symbols used, and which are actually transmitted.

This structure is received by the other correspondent and he first has to decode it. He has to decode the symbols he has received, but that is not enough. To understand the meaning of the communication, he has to compare the symbols with the object, with the image of the object implied, and this means that the second correspondent must also effect a process of reflection. The real object must be reflected in his thinking. Only then will it be possible to speak of a communication as having taken place.

Generally speaking, it must be added that each object of information implies or entails another aspect, a pragmatic aspect —actual action caused by a realization of what has been communicated. It could be said to represent action or a certain type of mental activity.

In the case of CETI, we have no opportunity for the receiving correspondent to reflect the contents and, as a result, we have only a symbol system. Strictly speaking we cannot distinguish a symbol system from a nonsymbol system (as any material structure), but that is a separate matter. Let us consider that we have some system about which we know only that it is a symbol system, but no more. We have here an example of an isolated symbol system. The question is: Can we understand what is implied by an isolated symbol system? Unfortunately, there are at least three fundamental objections to such a possibility.

The first, a very weighty objection, is the same kind of theorem which was mentioned by Idlis (p. 183), who implies that no isolated symbol system can interpret itself within a framework of a symbol system alone; it cannot explain the relationship between the symbols that make it up. This imposes a fundamental limit on all attempts to create interstellar languages such as Lincos.

The second fundamental objection, linked to the first, is that such a situation makes it inevitable that an isolated symbol system will reconstruct its own knowledge in the set of symbols used. It will observe a transplantation of its own knowledge— in turn an isolated symbol system. In other words, we are defining the symbols of that symbol system not from objective knowledge but on the basis of knowledge that we, ourselves, have. It is even a more difficult problem than a quite corresponding cybernetics problem, that of distinguishing the goal "for" and the goal "of" a self-organized system.

And thirdly, the structure of a symbol system cannot in any directed manner be linked to the meaning of the communication transmitted by these symbols. In other words, pure structure or code gives us no clue as to the real meaning of what is being communicated.

Also very important, to my mind, is the fact that in the process of such communication, even if they exchange roles in the case of a dialogue, both are interpreters; that is, each must interpret the range of objects that are discussed in the communication.

In our terrestrial practice, the following fact is of fundamental importance: the understanding of any symbol system requires that the symbols be repeated in situations we know but in different practical contexts; that is the only clue to decoding their message.

Very often in the CETI literature we hear it said that while an isolated symbol system cannot be understood, may not a situation develop in which simple contexts arise, associated with certain basic physical objects which are common to our system and to other extraterrestrial systems? Is it not possible to take advantage of that, to take advantage of the identical real object known to all and in this way decipher the meaning of a certain set of symbols, on the assumption that this is the subject discussed in the communication?

Generally speaking, according to the theory of cognition such a solution would be possible only if the real objects implied were the direct contents of our scientific knowledge. But this is not so, I must remind you. We can only consider this in a vulgarized reduction of the real state of affairs in the field of cognition. The fact is that the objects of the world around us are not direct contents of our human knowledge, and that is the whole trouble. The immediate contents of our knowledge is not the material properties or relationships that exist in the material world "itself" without cognition, but their reflection in our minds by means of ideal images which are the direct product of our practice; that is, those images whose nature unfortunately is not determined by any material phenomenon. This is the difficulty facing us. The process of cognition operates by ideal images. To have a direct relationship to reality involves some practical activity, and that practical activity alone enables man to come into contact with reality. For this reason practical activities are a basic expressed condition for cognition. Practical activity is what brings us into contact with the material world and this enables us to build up scientific theories. We cannot separate one from the other. Therefore, the cognition images we use in our scientific learning, the structure of that learning, the structure of our notion in symbols reflecting the real world around us—all this includes as the objective properties of the things around us the instrument of our learning.

The cognition conditions are, unfortunately, inseparable from the objective properties, and it would be naïve to think that in terms of conditions in our theories we can always distinguish the objective properties of a phenomenon from this fundamental instrumental aspect of the matter.

All this brings me to the conclusion that in order to understand the set of symbols used by another civilization, there must be observed a severe condition. What is needed is a close identity of the historical background of the two societies. That identity must be so great that to some extent we must speak of a second earth or "earth prime." That I feel imposes a very stringent limitation upon the number of possible civilizations with which we can communicate. Semantic communication is extremely limited in view of these difficulties, although the problem is open to further discussion.*

KUZNETZOV: I would like to say briefly something about the exchange of information between civilizations in the absence of any a priori understanding about coding methods. For analyzing the problem, it is necessary to define the terms *communication, coding, signal,* and *modulation.*

A communication to be transmitted is an entity of information, a concept about some process or about a relationship among phenomena. A signal is a communication converted into a form convenient for transmission. Coding is a method of conversion into a signal. Modulation is a change in the parameters of emission serving as a carrier of the signal.

* [Editor's note: For different views, see pp. 335, 346, 347, and preceding discussion].

We want to receive or transmit communications, but we can only exchange signals. This presents a number of difficulties. If the distance between the correspondents is large, the transmission will evidently take a long time. If one correspondent speaks to the other without listening to the reply, this makes things difficult. There is practically no a priori information about the coding method and we can only count on our correspondent being clever enough to understand what we are saying. But it is precisely this contact with intelligent beings that is intended. As for the channels to be used, evidently these will be varied, and this is not a question of principle.

Now about the mode of coding: in general, the mode of communication between two civilizations must be such that the main strategy must be to further the search. One correspondent tries to understand what the other will do.

Now, how can we transmit notions? This can be done by models. (The fullest model, of course, is the object itself. It carries all the information.) Therefore, we must transmit models by means of signals. Models at different levels are possible. Models can be classified immediately as those that retain most of the information about the object, about the structure of the communication, and others that are more abstract.

A second type of communication is artificial language of the Lincos type. There must be learning because structures cannot be transmitted. As an example of the first type, let us assume that we want to transmit a drawing of a cat. We have a set of coordinates. We can transmit this picture in the form of a signal that will be coded in natural coordinates. The natural coordinates in an exchange of information will be tallied evenly at the transmission and the receiving end only for evidence of the signal. It will be expedient to use, say, the parameters on which the search for the signal is conducted. All of the correspondents will, say, be scanning frequencies. One geometrical coordinate that can be used is time; another is frequency. That means there are two searches, and they yield a two-dimensional picture of the cat (Figure 41).

In this way, we can transmit an informational signal without even special call signals. If we repeat that picture many times, in any frequency, we will have a recurrent communication. If the search is conducted in a narrow frequency range, the picture will be transmitted repeatedly and at these moments in time there will be impulses and with the pictures repeated again and again we get a recurrent system of impulses which then become a signal to attract attention.

The instrumentation for receiving the signal will at the same time carry out search and analysis, and since it will be sufficiently comprehensive, it will receive a large number of parameters and can be used for ordinary radio astronomical activity. Since the pursuit of the ETI search must take into account a large number of parameters, it will be suitable for radio astronomical work, too.

The CETI problem is a multifaceted one. There are people here from different branches of science. There must be good coordination in working on the various aspects of the problem. We must have a rational distribution of effort.

PETROVICH: Referring to the first part of your statement, do you consider it possible to transmit individual ideas?

KUZNETZOV: Yes, by means of the model it is convenient to transmit pictures—pictures that will correspond to certain ideas.

PETROVICH: So, there is a system; one can only understand an idea in a set of ideas?

KUZNETZOV: Yes. Of course, a certain measure of learning is essential.

BRAUDE: You are transmitting a picture but that same picture will be received in another world and interpreted in a certain way. Don't you think that it may happen that in another world

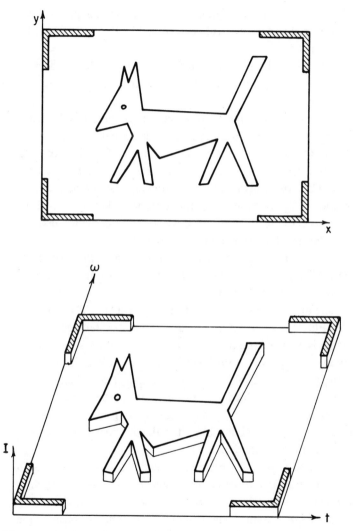

Figure 41. Transmission—by using intensity, frequency, and time coordinates—of the image of a cat.

with creatures with a different refractive index, the pictures will
be distorted and the cat will be nothing like the cat because the
medium is different?

KUZNETZOV: The picture would be distorted; distorted cats
would be received. But the topological structure of the picture
is the same.

SUKHOTIN: I would like to comment on decoding problems and
specifically on distinguishing an artificial signal from a natural
signal. The importance of decoding methods for CETI problems
may be argued. Professor Freudenthal proposed a teaching program (Lincos) which reduces the role of these methods to a
minimum. However, even a Lincos message does not lend itself
to easy decoding if it is received without an instruction text. On
the other hand, it is sometimes asserted that it is impossible to
understand messages from ETI. This remark holds true, of
course, for a message which is too brief with regard to the alphabet or deliberately confused.

There are two aspects of the decoding problem: distinguishing
an artificial signal from a natural one, and its interpretation in
case it is artificial. Let us start with the second question. Decoding is not a single problem. It can be broken down into
many particular problems that are complicated in themselves.

The decoding problems are grouped into three classes, three
stages of the decoding. I shall abstain here from discussing the
decoding of pictorial messages. I shall confine myself to languages of the human type.

The three stages here are grammatical analysis, semantic analysis, and translation. At the first stage, one has to solve such
problems as identifying morphemes (that is, the smallest meaningful linguistic units), singling out words, and establishing the
structure of words—that is, revealing the relationships between
the morphemes, discovering the syntactic classification of words,
establishing the boundaries between sentences, and so forth.

The second stage calls for establishing the semantic equivalents of words and expressions, singling out elementary situations and establishing the structure of the text; that is, revealing the relationships in the set of elementary situations.

The third stage consists in the mapping of the language of the message into the language of the investigator. This mapping is a formal analogue of translation.

If one deals with human languages there is a fourth stage of decoding consisting in the reconstruction of pronunciation.

The main instrument of linguistic decoding is the decoding algorithm. By decoding algorithm I mean a system of three objects: First, the set of admissible interpretations (solutions); second, the function of quality, estimating each of the admissible interpretations—this function is calculated from the text under consideration; and finally, the third object is the computer procedure or the algorithm proper for finding the maximum or minimum of the quality function. For some simple problems, we have algorithms tested experimentally in computers, including some experiments that have not been mentioned in the volume edited by Kaplan.* I shall give an example of a simple algorithm to make it possible to tackle the first problem, that is, distinguishing an artificial message from a natural one.

This algorithm solves the simplest classification problem, the classification of some alphabet of linguistic units into two classes. If the alphabet is an ordinary list of letters, the algorithm simply determines which letter is a vowel and which a consonant. The set of admissible solutions in this case amounts to a number of divisions of the alphabet into two subsets which do not intersect. The quality of the division is determined by the following hypothesis: In any text in a natural language the sequences

* B. V. Sukhotin, in S. A. Kaplan, ed., *Extraterrestrial Civilizations: Problems of Interstellar Communication* (in Russian). Moscow, 1969.

vowel + consonant (VC), or consonant + vowel (CV) are more frequent than the sequences VV or CC. The number of sequences VC and CV (for a given text) depends exclusively on the division. In this way, that number may serve as an estimate of the quality of the division.

This algorithm has been tested repeatedly and has yielded good results. There is a computing procedure that defines the maximum of this function.

Now we can pass to the method of distinguishing an artificial signal from a natural one. I consider that a message sent by ETI is in some respects intended for decoding. What does this mean from the point of view of decoding methods? It means that in analyzing this or that linguistic phenomenon, in solving this or that decoding problem, a correct solution must stand out against the background of inadmissible solutions. This, in its turn, means that in the set of admissible solutions there must be not only very good solutions but also very bad ones, and the degree of intelligibility of a message may be determined by the difference between the quality of the best and the worst solutions, the greater difference corresponding to the greater degree of intelligibility.

A linguistic phenomenon for which this difference is large can be termed diagnostic. If one intends to process a random sequence of letters by means of the algorithm that discovers vowels and consonants and to determine the estimate, evidently that estimate will be close to zero. The situation will be the same for a strictly periodic text where the vowels follow upon the consonants strictly. For this type of texts (random and strictly periodic), vowels and consonants are not diagnostic. Such texts are not intended for deciphering. We may assume that in a text intended for deciphering, at any rate, some of the linguistic phenomena will be diagnostic. Professor Braude had suggested, by the way, that the proposed estimate is similar to the signal-to-noise ratio which is used so widely. In this way, if we investigate a message with the help of a certain deciphering algorithm, we shall not only get some interpretation but we shall also be able to estimate the validity of our analysis; the more such attempts result favorably, the greater the chances that we are dealing with a rational message. It is important to note that this result may be obtained without going into intricate semantic analyses by means of the simpler grammatical analysis.

PETROVICH: You spoke of intelligibility. Is that in the same sense as used in CETI, this predictability of the regularities with which the symbols occur?

SUKHOTIN: At the beginning of my talk I did not speak of the intelligibility terminologically. Later, however, I tried to define this notion. I believe that a message intended for deciphering can be understood, whereas a message that is not intended for deciphering evidently cannot.

PETROVICH: Could you estimate the time needed for learning Lincos, considering the large number of exercises that have to be done by the student?

SUKHOTIN: You see, I am studying methods that make an instruction program unnecessary. I do not count on getting a message of the Lincos type.

OLIVER: I don't think they are going to talk in vowels and consonants. I think they are going to talk in binary clicks.

SUKHOTIN: I, too, do not think they will use vowels and consonants. I mentioned the problem of discovering vowels and consonants for two reasons: first, it is a good example of a simple decoding algorithm of the classification type; second, on the basis of this algorithm I introduced the important idea of intelligibility.

TOWNES: I wonder if anyone would comment on the possibility of a space missile being sent to us with a complete dictionary based on pictograms, and then lots of literature with it. If one assumes some length of time allowed for travel, it would seem to me a rather efficient way for communication which completely eliminates much of the uncertainty we have discussed.

MORRISON: This particular suggestion has been studied by Doctor Bracewell, who, in a paper some time ago, discussed the question of probes. This summer he prepared a second paper in which his considerations lead him to conclude that for a very large number of possible channels, or for a rather small number, such a method is not efficient; but there exists an intermediate range in which his studies at least seem to indicate probes to be the method of choice. I think very likely it should be put into the box for later consideration.

MINSKY: I wonder if many of you have read Hoyle's novel "A for Andromeda." The idea is that instead of sending a very difficult-to-decode educational message of the kind that Freudenthal describes, and instead of sending a picture of a cat, there is one area in which we can send the cat itself. Doctor Kuznetzov mentioned that there is one property that is independent of distortion and that is topology. Topology is that which remains after distortion.

Briefly, the idea is that we can transmit computers. A computer is an absolute topological device and if we send pictures —and I prefer the proposal of Drake for television in prime number format rather than this frequency modulated form of television—we can send a series of Drake pictures which explain how to build a program or, rather, a computer with the program. Then our ETI's can run this computer and the computer will contain the teaching program of Freudenthal, except in a form that can interact with the people and learn their language. You can transmit computers without fear of any ambiguous interpretation. I believe by sending Boolean logic diagrams, one avoids most of the possible ambiguities.

I would like to convince you that it is possible to make a computer program that could interact with someone else's language in a useful way. I will read you a few sentences of man/machine interaction from a new computer program completed about a year ago by Terry Winograd in a doctoral thesis at the Massachusetts Institute of Technology. This program deals with a small mechanical world in which there are only a few concepts: the concept of a solid body, the idea of one body supporting another, the idea that something can be in a box, that something can be to the right or to the left—Cartesian coordinates, and things like that. The program and a human are discussing this mechanical world and the person asks, in English, typewritten: "What does the box contain?"

The machine answers, "The blue pyramid and the blue box." [This is a world of children's toys.] "How many blocks are not in the box?" "Four of them," the machine says. "Is at least one of them narrower than the one which I told you to pick up?" "Yes, the red cube." Well, it goes on this way. At one point the operator says: "Is there a steeple?" The machine says, "Sorry, I don't know the word 'steeple.'" The man says, "A steeple is a stack which contains two green cubes and a pyramid, for example." "I understand," the machine says. "Are there any steeples now?" "No," the machine says. "Build one," the man says, and the machine does. It reaches its mechanical hand out and moves the cubes around and builds the structure that has just been described. In another part of the dialogue, one asks the machine, "When did you pick up the pyramid?"

The machine replies, "While I was stacking up the red cube, the large red block and the large green cube." "Why?" "To get rid of it," the machine answered. "Why did you do that?" "To clear off the red cube." The man asks, "Why did you clear off that cube?" The machine answers, "To put it on the large green cube."

The point is that this machine not only takes instructions in

English, in natural language, but it can be asked why it obeyed instructions and it will give a hierarchy of mechanical reasons until finally, if there is no other reason, it must reply, "Because you asked me to."

So it appears that in 1970 we crossed a threshold of being able to deal with semantics in computers. Over the next ten or fifty or one hundred years, I am quite confident that we will be able to write computer programs that are reasonably intelligent. Then most of the problems of decoding that Freudenthal faced, not entirely successfully, I think, will begin to disappear.

SAGAN: If we build a computer as instructed in a radio communication from ETI, would we have established a communications channel with a much higher bit rate than if we were left to our own devices (so to say)? However, the intentions of the ETI transmitting computer instructions in the story Doctor Minsky mentioned were not entirely benevolent. As Morrison has stressed, we would construct such a device only when we understood it thoroughly.

CRICK: Can I ask Doctor Minsky the length of the program to which he referred?

MINSKY: The program of Winograd is very large, some 10^6 bits. There is no way to estimate how much smaller it can be made. It is conceivable that 10^4 or 10^5 bits will do.

I would like to add one point, though. The real heart of this proposal is that although it is hard to transmit some isolated concepts, concepts about processes, and particularly concepts about digital processes, are exceedingly easy to transmit—because as soon as the correct interpretation of the first symbols is made then the ETI can simulate and run this computer more or less immediately. It requires no other concepts because ETI are essentially building it for themselves and experimenting with it to see what it will do. In the Freudenthal situation, one has to tell him about a mathematical concept and then tell him more and more and more about it; I feel that process never ends.

KUZNETZOV: I would like to ask Doctor Minsky this. I was somewhat taken aback by his reference to the cat. Of course, a living cat is better than a picture of a cat but the question is, How are we to transmit that developing system of a living cat? So the same problem here is simply transferred elsewhere, but the idea is a very good one.

SAGAN: You transmit the genetic code of the cat.

KUZNETZOV: That is also a very good idea but how are you going to do that?

BRAUDE: I would like to ask Doctor Minsky whether he took into consideration in these programs the fact that the transmission of signals passes through a dispersing medium where the topology may not be preserved?

MINSKY: No, we sent simply the sequence of pulses. But there is no dispersing medium which can change the order in which pulses are received, although dispersion can certainly spread that out so that they are very difficult to preserve.

OLIVER: I think the suggestion of Doctor Minsky is a fascinating one. It raises an interesting question in my mind. Is the information transmitting capability of a computer that has been described by transmission over the channel greater than the information required to describe the computer? If so, we seem to have violated Shannon's limit.

MINSKY: No, the information is the same. The point is the character of what we transmit. I claim that most information that people have that is important is not facts but processes. In particular the process by which you parse a language and understand a grammar is much more important than the grammar itself. The reason Winograd's program is more successful than any previous program in dealing with the structure of language is because in Winograd's program the structure of language is described as a process, that is, the process by which you decode language, rather than by the traditional methods of Chomsky and others in which one attempts to describe the structure of language as a sequence of isolated rules. I think Shannon would agree that you get much more for your bit if it is a profound bit of knowledge than a trivial one that is transmitted.

THE CONSEQUENCES OF CONTACT

MORRISON: One subtitle for my talk might be "The art of prophecy." The best thing I can say about that art is that it is now a lost art, if indeed it ever existed. We are not given such complex insights. We have organized this discussion on the social implications of contact with extraterrestrial intelligence, not because we expect to say very penetrating things about it, but because we need from the beginning to accept responsibility for the effect of what we may be planning. It is more as a sign of that responsibility than as a contributor to prophecy that I am here.

In order to discuss the conceivable effects of contact I think I must give some model of what happens once we have detected ETI. I believe that, first on one channel, and, with the passage of time, on increasingly many channels, we will receive a signal with the structure shown in Figure 42. Time is increasing to the right. As yet I give no scale. The signal is divided in many parts in the following time pattern: Each block letter will indicate an identical portion, A identical with A, B with B, and so on. In between the A blocks, another set of identical B portions each, of

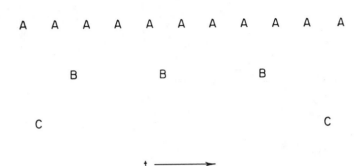

Figure 42. Schematic representation of the structure of anticipated received signals from an extraterrestrial civilization. An acquisition signal A which is readily recognizable is transmitted frequently. With less frequency are transmitted longer and more difficult parts of the message B, C, ...

course, distinct from A are repeated, less frequently. Then another repeated pattern, C, perhaps not quite so rapidly repetitive, occurs. The figure shows some estimate of the time in the channel, small times for A, larger times for B, still larger times for C. There may be more structure yet, but after a small list A,B,C, . . . then all other parts will have completely nonrepeating signals. The signal originator must maintain interest in his signal by giving the very simplest part of it in a very rapidly repeating sequence; this is what I call the acquisition signal, A.

There are many radio stations in which a time signal or the main call letters are broadcast every hour. One may well imagine —on a time scale which I cannot easily guess here, on an unspecified time scale, but not longer than 10^6 seconds—such brief, simple, probably wide band and high intensity signals, meant only to call attention to the existence of a message. As I tried to indicate earlier, we must sharply distinguish an acquisition signal from an efficient message.

In between, I imagine a signal, B, much more complex than A but less complex than the ultimate message, which is, in fact, a decoding signal. Here, in particular, I wish I had time to go into the discussion of Doctor Panovkin (p. 320), but I take the view that the signal itself, by its physical nature, by its intention, by other features held in common between transmitter and receiver civilizations, may be the way around the logical difficulties of which Panovkin speaks.

On that assumption, I imagine a third structure, C, to the hierarchical signal, which is essentially a language lesson. Here I quite agree the decoding is a map of the symbols into some referents. The language lesson must include much more buildup of context, the rest of the time occupied by an extremely complex structure, probably taking full advantage of bandwidth and of ingenious modulation, and the whole thing occupying a channel—who can guess how wide?—but I would say not greater than 100 megacycles in bandwidth. That is a very wide band.

Now, this complex structure arrives at our receiver; it is not something like a story in the daily newspaper. It is the object of intense socially required study for a long period of time. I regard it as much more like the enterprise of history or science than like the enterprise of reading an ordinary message. Obviously, there is lots of storage implied, tape upon tape upon tape, and, as usual, the data will far exceed our reading ability! The data rate will for a long time exceed our ability to interpret it.

I therefore thought it useful to look at the scale of this signal. Can such a signal quantitatively approach a large part of human experience? It is dangerous to judge such events, obviously, but I came to the conclusion that it will not be a major quantitative addition to human experience. I estimate human experience (and I will be glad to discuss these estimates with other people) at 10^{20} to 10^{23} bits (let me not try to claim such accuracy; I have given myself three orders of magnitude, but that is only the middle of the guess), but I think you see that we have been talking logarithms by orders of magnitude. Perhaps human experience, total human experience, amounts to one mole of bits.

Most human experience remains unexpressed. It is the summed content of the conscious and unconscious mind of all humans. It is translated into public experience by the behavior of human beings, one with the other—by conversation, writing, craftsmen's work, making bread, whatever—I think we lose a great deal before transfer, and the number may go down to 10^{16} or 10^{18}. The present rate of human experience approaches 10^{11} bits per second, or 10^{18} bits per year.

We all know that when we look at big numbers in an unfamiliar context, some sensible unit must be introduced. As a guide, I suggest for the discussion the use of all that we know of Hellenic culture. No one can deny the importance this subject has had for two or three millennia over the whole world. It corresponds to a one-way message with 10^9 bits of text.

It is unfair to refer to the text alone; the context is very important. I make a very crude estimate of the number of volumes of photographs, etc., necessary to describe the architecture, climate, pottery, fish, botany, etc.—whatever gives us the context of Hellenic time. I think it is not much more than 10^{11} or 10^{12} bits.

So you see that we can learn a great deal and experience a major effect on the formation of our own mind and institutions, both individual and social, with a number of bits which is much smaller than the total human experience, by a factor of 10^{10} or 10^{12}. A channel at 10^{18} bits per second, a large estimate, or even a hundred such channels, would be much more in total than Greek culture after one year, but would remain small in quantity compared to our own human experience. It looks to me as if it would take ten thousand years at that rate, roughly speaking, for a fast channel to add up to what we humans have already— and of course in ten years we would have generated a similar amount more. This crude quantitative discussion tells very little of this wonderfully complex cosmos, but it establishes for me that it is the quality of what comes over, and not the quantity, which is important. Of course, the very acquisition signal is the most important step of all. It is entirely possible that the proper receiving strategy will involve looking for these acquisition signals in a different way from a message channel. I hope I have conveyed the idea that it is not reading some simple telegraphic message like a newspaper which we can anticipate.

The recognition of the signal is the great event, but the interpretation of the signal will be a social task comparable to that of a very large discipline, or branch of learning. In that light, I think that a sober study will show that a message channel cannot open us to the sort of impact which we have often seen in history once contact is opened between two societies at very different levels of advance. There will be absent across space, of course, *any* military dominance, whether as in Mexico in the sixteenth century when military dominance from the outside was dependent upon a local alliance, or as in the Canary Islands or in Peru where it was fully external. Nor will there be the thrust of any technical economic competition, like that which induced famine among the highly developed Bengal hand weavers, faced with the machine-made cloth of Manchester. At most, we expose ourselves to the dangers or opportunities faced by the Japanese society on two occasions in its history, once when it encountered the enormously strong culture of the T'ang through a very few persons traveling; or in the nineteenth century when the Japanese system changed entirely after what was only a threatened invasion, but a threat which brought out internal strains that were very deep in Japanese society. So I am confident that on this kind of model, which seems to me very plausible, we could imagine the signal to have great impact—but slowly and soberly mediated, transmitted through all those filter devices of scholars who have to interpret and publish a book, and so forth. Note that the total gloss on Greek thought is at least as voluminous as the Greek texts themselves!

Considering two-way acquisition to be very rare, the communication is necessarily between societies, not between persons. I think the Greek analogy is again very valuable here. The content of the signal will only in part answer the questions even of the natural scientists because of the limitations of uncertain contexts and different materials. It will not begin a Golden Age in which the meaning of some communication about new kinds of elementary particles is immediately obvious. Physicists will have to decide whether to study this body of knowledge, in which the answer is surely contained, or to do the experiment for themselves. (Already it has reached that point in our own scientific literature in many ways.) I have a high opinion of the ability of ETI; still I wish I could believe that all ancient philosophical and theological questions would be settled at once and to everyone's satisfaction by the first messages—but experience leads me to think that will not happen.

Those who wish to understand the message will see the question settled; those who wish to reject the message will certainly reject it, since its guarantee of veracity is not in itself more than that of any other text.

Most of this very complex signal will contain not mainly science and mathematics but mostly what we would call art and history. For me, it is plain on combinatorial grounds that our society or any long-lasting society will solve many scientific and mathematical problems more easily for itself than by studying tape recorded from extraterrestrial sources. But we are not at all able to reconstruct the imaginative, the fictional or the historical events of a distant future. The possibilities are too great. We have no clues about how to do it. There are many more folk tales than there are laws of mechanics. That, I think, will give the fundamental, the lasting motive for such a channel. It will produce these novelties for us as a social obligation felt by the transmitters, who for their part in the past had already received many such novelties.

I imagine that the most important reason to search for such a channel is that if it exists—and that of course is the most important question from where we stand now—it will produce a view of the universe less alienating than the one now characterisitc of the general population of industrial societies. I do not speak here of practical help, which of course would exist but which is probably more difficult and in the long run less important.

Even taking a signal model which is pretty optimistic, it still turns out quantitatively that such a signal will never come to dominate all of human experience but will nevertheless be a very rich contribution—equal to that of Greek culture in under one channel year—which cannot be studied in the style of reading a newspaper, as Athenian life and thought cannot, but only by careful investigation, with the full equipment of cooperative human scholarship over an extended period of time, a discipline rather than a headline or an oracle.

PLATT: It seems to me that Morrison is saying that the medium is the message.

SHKLOVSKY: Is it a joke?

PLATT: No, no joke.

SHKLOVSKY: Would you rephrase it?

PLATT: The medium is television. It brings the message that there are television stations.

MORRISON: The existence of the signal is indeed its most important property.

LEE: Do you think that the signal will contain any questions?

MORRISON: I do. I think the signal will contain many questions. It will contain, of course, proposals for the construction of much better transmitters, but the time scale will be very slow. I imagine the characteristic distance to the transmitter is tens or hundreds of light years, so they don't expect an answer before tens or hundreds of years time.

LEE: Can you give us an estimate of the length of time it will take to transmit the main message?

MORRISON: I think the message is effectively infinite in length, that it will last the full time L, whatever that time is. They will keep constantly sending what they know, or rather carefully selected pieces of what they know. What would you do?

LEE: Does that mean we tune in at some arbitrary time?

MORRISON: Of course; what you get is a hierarchical message, rather like going to school; finally you learn enough to read the latest *Physical Review,* with its new material sent out every once in a while.

DRAKE: It seems to me unfair just to use the number of bits as a measure, even with Greek civilization. The first complaint is that the bit rate for human civilization is for bits which are almost entirely optional and have no lasting significance at all, which is to say some bits are better than other bits. For example, perhaps for only 10,000 bits there would come the design of a tachyon telescope which would have vastly more impact on civilization than all of the bits from Greece.

MORRISON: I disagree strongly with that implication. Ten thousand bits is 2000 words. To find a paper in *The Physical Review* that is that good and that brief, you will have to wait a hundred years.

DRAKE: Maybe tachyon telescopes are easier to build than you imagine.

MORRISON: Yes, they would be quite useful, if they occur.

PANOVKIN: I disagree with the point of view that mathematics are universal. What are the sources of mathematics, the bases of mathematics? There may be different axiomatic bases of mathematics. Such phenomena as infinity are generalizations from human knowledge. Other societies may have other generalizations.

BRAUDE: Do you think that some of the information contained in the messages will have meanings which are beyond human experience?

MORRISON: Surely, yes. The number of bits has nothing to do with it—it is just a number. It has nothing to do with what the content is. The message can be very rich; I say only if it is very far beyond your experience it is hard to come to understand.

Let me just answer Drake, which may also help answer the last question. Galileo saw the telescope when one first came to Venice in 1608. In the year he made his first telescope, they were selling them as novelties in the streets of Paris and Amsterdam. They were 3-power. When you looked through them, there was a distorted mess. They were mere toys. Galileo made hundreds of lenses in his lifetime. With four that he carefully selected, he discovered the moons of Jupiter and initiated telescopic science. He had to develop lens-making, not copy the bad lenses.

DRAKE: It seems to me you are agreeing with me, but let me give another example. How many bits does it take to transmit $E = mc^2$?

MORRISON: In the year 1600 A.D., say, there you are studying Fibonacci series, and an angel appears to you and writes $E = mc^2$ in your notebook. You can't even ask what it means. Does that help you much? It took the scientists of Europe ten years to understand Einstein's 1905 formulation, and that was very much more in context. I imagine it was the richest single physics document we will ever read—I agree with that entirely —but I don't think it was simple.

BURKE: I wish to address a remark or question to Professor Morrison concerning the estimate of the approximate time that would be needed to give us a reasonable chunk of—what was it?

—10^9 or 10^{10} bits. The part that would worry me in this estimate is the fact that the uncertainty is in the logarithm. If all these bits arrived on a 10-year time scale, it could indeed be very important.

MORRISON: Please don't misunderstand. In the first place, you have some numbers wrong. What I said was that the human experience would be repeated only in ten thousand years. In my crude estimate, the Greek-culture number might be reached in 100 seconds; I have deliberately assumed a very wide channel. Working out the message would take very much longer than that; you are accumulating data much faster than you are able to interpret it.

OLIVER: I agree with Professor Morrison's remarks that there will be no culture shock from detecting the first signal, which I think would be a beacon signal. But I would like to point out that after the round-trip light time has elapsed and the two cultures know they are in contact, there is no technical difficulty whatever in getting information rates of 10^7 or 10^8 bits per second over 500 to 1000 light years.

MORRISON: I assumed 10^9 bits per second to begin with, so I am still overestimating by a large factor.

MCNEILL: I have been listening to Doctor Morrison's remarks with some special concern to see whether any of the skepticism with which I approach the whole notion of extraterrestrial communication is relieved, and I regret to report that that is not the case. I am more deeply impressed than perhaps Professor Morrison is by the difficulty of deciphering the message. It seems to me that any reasonable assumption of the difference between earth life and the life on planet X, even though planet X had technical mastery comparable to our own or superior, would be so great as to make mutual intelligibility a very strained and not to be hoped for probability. Our intelligence, unless I misunderstand, is very much a prisoner of words, a prisoner of language, and I don't see how we can assume that the language of another intelligent community would have very many points of contact with our own. The differences in biochemistry, in the sensory span of intelligent beings, sensibility, in such things as size of body and neuron pattern, all seem to me to add up to a high improbability of mutual intelligibility.

But let us suppose that those difficulties prove in fact soluble, that you can not only receive the message but can decipher it; then it seems to me there are two other considerations that I would put great weight upon. One is that if communication should be opened with a technologically superior civilization, that civilization might choose to exploit the earth rather than tell us fairy tales and history.

The only parallel that one can draw upon is the history of human contacts on Earth and there, unless my reading of history is radically wrong, contact between men has shown that those who had the power used it. Athens was a brutal master.

MORRISON: Oh, no—Socrates and Plato were in no way brutal masters against fifteenth-century Europe.

MCNEILL: What do you think Pericles was doing?

MORRISON: You misunderstood me. He was a tyrant in life; but once in Thucydides and dead 2000 years he was only a moral example!

MCNEILL: Suppose now that we can eliminate the danger of being exploited, the danger of invasion or control from a distance, so that we human beings can neglect what we choose to neglect in the message and pay attention to what we choose to pay attention to; then I think, assuming we can decipher, assuming that there is no danger of being exploited, then there would be a clear net gain to human knowledge. If we can be assured of immunity against exploitation and we can decipher, both of which I remain skeptical of, then the gain is real.

There have been cases in the human past when very remarkable consequences followed from the opening of new communication nets. The one that comes most readily to my mind is the consequence of opening regular communication among the scientists and interested gentlemen of the Royal Society of England with other similar societies that quickly sprang up in other states of Europe; this was one of the prerequisites, I think, for the development and firm growth of modern science. But I would point out that that net fell, as it were, upon a fertile field. There were people who shared a very great deal in Holland, in England, in France and Italy, and it is this common ground which I think we will lack with extraterrestrial intelligence. So I find it very hard to believe that useful scientific information, historical information, literary experience, can move across that cultural barrier.

Now, I do not deny, nor do I doubt, that should communication be opened, the reaction among mankind would be very strong—not because of the content of the message but simply from the fact that a message could in fact be received. Such an experience would say to us human beings "We are not alone in the universe." And this, I think, has powerful psychological reassurance value and, should it be achieved, might by itself quite justify any expenditures made in the search for extraterrestrial intelligence.

I must say that in listening to the discussion these last days, I feel I detect what might be called a pseudo or scientific religion. I do not mean this as a condemnatory phrase. Faith and hope and trust have been very important factors in human life and it is not wrong to cling to these and pursue such faith. But I remain, I fear, an agnostic, not only in traditional religion but also in this new one.

LEE: You say you are an agnostic in both religions, but the existence of which god are you putting your money on?

MCNEILL: Well, is that a necessary choice? I did not mean you shouldn't try. I do mean that as far as my judgment goes, the expectation of large payoff in human time, in a thousand years, seems to me highly improbable.

DRAKE: Do you know of any other agnostic sect with missionaries?

MCNEILL: Yes, Buddhists were agnostics, or at least they were atheistic. Perhaps "agnostic" is the wrong word to use about Buddhists, but they were missionaries. I must say that secular religions are not new. There are a great many of them and many of them have missionaries.

VON HOERNER: I think if a Stone Age culture comes into contact with us, this means absolutely the end of that Stone Age culture, although this may take a while; and if we come in contact with some superior civilization, this again would mean the end of our civilization, although also that might take a while. Our period of culture would be finished and we would merge into a larger interstellar culture.

MCNEILL: If I may reply, I think the word "contact" hides two quite different realities. Contact which leaves the human race free to accept or reject, which takes the form of radio signals or something of that sort, does not necessarily, or even probably, have a very drastically destructive effect upon it. It is when we have no choice, when some superior power impinges upon us, not simply by delivering intelligible symbols but in some other fashion, that the end of human civilization as we have known it would then become an expected consequence.

VON HOERNER: I would think the means of communication, whether by force or only by signals, defines only the time scale of the process.

MCNEILL: That assumes the message is intelligible and we choose to do something about it—that we can tell them we understand.

VON HOERNER: Given enough time, it will be intelligible.

MCNEILL: But they may not want to be understood.

CRICK: I would like to make two comments on what you have

said. I think you do not allow for the existence of mathematics. This is a natural language which it can be argued would be common to both parties. Moreover, it is easily put into a form, into a notation which is easily transmitted.

For the same reasons, it is not too difficult to start explaining elementary physics, and then to proceed in easy stages to more complicated physics, and then by further stages into physics we do not know. This is common knowledge. I understood Professor Morrison to say that when you come to something like literature, communication would be extremely difficult. To that degree, we agree with you.

MCNEILL: May I respond briefly? The confidence that I know many mathematicians and natural scientists have that they have a universal language seems to me a case of chauvinism, to use our favorite term. I can't prove it, but I don't think you are justified in assuming automatically that our mathematics is commensurate with their mathematics.

CRICK: But you agree that it is necessary for them to have mathematics in order to make a communication?

MCNEILL: I should think so, yes.

CRICK: And physics, too?

MCNEILL: I should think so.

CRICK: And I would not express an opinion on the point when so many distinguished mathematicians are here, but I would be surprised if they did not agree with me rather than with you.

MCNEILL: Oh, I know that.

CRICK: I don't think you realize the force of the argument.

My second point is a matter of human psychology. You are quite right that in this room there is a mixture of believers and skeptics, and you have correctly recognized the believers, but some of us are also skeptics. However, the history of science shows that not only are believers common but that skeptics are too skeptical. It was only about 100 or 150 years ago that

Auguste Comte said we shall never know of what the stars are made, and it was only 45 years ago that Lord Rutherford said that atomic energy was impossible; so I must caution you about your figure of one thousand years.

MCNEILL: Well, I certainly agree that most skeptics have always been very skeptical and believers have frequently been the carriers of major innovations. A man like Kepler was a crackpot, but he nevertheless got a long way in figuring out the planetary orbits.

OLIVER: I found Doctor McNeill's remarks very interesting because they reflect opinions that we are apt to encounter in the educated person who is not intimately acquainted with science and with the problem of interstellar communication. I say that they have, then, political significance.

I would like to amplify Doctor Crick's remarks slightly. I think the idea that we would have a great deal of difficulty communicating rests on an underestimation of the amount of knowledge that we would automatically have in common with any other intelligent being or race. I would remind Doctor McNeill that another intelligent species very likely also has eyesight and that it is technically possible to send messages which are very obviously decodable into pictures and that, therefore, the avenue of teaching is open to us that is open to any two cultures that do not understand each other's language. We can draw pictures.

Secondly, I feel that Doctor McNeill overestimates the ease of interstellar travel. I would not be one to join the skeptics that Doctor Crick has pointed out who have made mistakes in the past by saying it is impossible, but I would say it is many orders of magnitude more difficult than communication by radio, and that any civilization that had achieved interstellar travel would be so far advanced as not to bother with us.

PEŠEK: My personal opinion is that CETI can help us to solve our terrestrial problems and increase the lifetime of our civilization, that is, increase L in equation (1). The possible consequences will be a revolution in science and technology.

GINZBURG: What I would say is along the same lines that Doctor Oliver mentions, but I feel that it is a very important point. You see, we definitely have here believers and skeptics; but, it seems to me, there can also be obscurantists—people who say that there is no problem at all, and therefore nothing to think about, nothing to do. From this point of view there is nothing more dangerous than to speak about the danger of communication. I do not, of course, see anything in Doctor McNeill's remarks in common with this. But I mean if we would like not to close off the problem altogether, forever, we must stress with the utmost positiveness (and I believe we can do this) that there is absolutely no danger. First of all, there is no danger when we only receive a communication. If we do not answer, they would not do anything. They wouldn't even know that we have received the message. I cannot even imagine any danger because they can't understand that we are looking for their signals.

But even, as was mentioned by Doctor Oliver (and I agree), even from the point of view that we try to answer, even in this case it is extraordinarily improbable that there is any danger. We all, independently of our views, are interested in the CETI problem. Our obligation is, I feel, to stress that, in any sensible way, this problem has no danger for human society. I believe we can give a full guarantee of this.

SHKLOVSKY: I would like to quote the words of the well-known Professor Sakharov, that such information would be useful for sensible and kind men and would be dangerous for silly and rude men.

MORRISON: I think this has been a very interesting discussion. Many of the points which I would like to make I think have already been made. Let me repeat them.

No mere symbol string, no set of mathematical symbols alone

would be the content of the message, but a very rich three-dimensional, moving, carefully scaled cinema. I think this would go far to establish a context on which experience would play. I think the problem of universality of mathematics is mostly a misunderstanding of language. A mathematician quite correctly says: The key to mathematics is infinity. But I imagine that the principal mathematical basis of the unity we seek, of the communication, would be in the beginning, restricted only to the finite mathematics of integer numbers, and these, I think, are likely to be the most universal structures.

I quite agree that the refinements of logic will be just as hard to establish from the distant message as they are here, but I really think the richness of signaling, assuming a wide radio-frequency channel and extraordinarily careful planning of the message, would be, as I tried to emphasize, no easy task to decipher, but fully possible—not as one reads the newspaper but as one works out a rich, difficult textbook of an advanced subject full of diagrams, hints, and examples.

RETROSPECT

SAGAN: I believe I can speak for all of us in saying that this has been a remarkably stimulating and productive conference. A search for extraterrestrial intelligence can appropriately only be international and this conference has had a strong international character. A search for ETI requires the widest range of scientific disciplines and I believe there have been few conferences with expert representation from as broad a range of subjects as at this meeting. We have of course not come to a solution of many of the questions regarding search strategy. But there has been a range of important new ideas expressed here as well as the first public announcement of a small but very important on-going search program. I believe that in the long run the significance of this Byurakan conference will be that it made the subject of communication with extraterrestrial intelligence scientifically respectable. For the American delegation may I say how grateful we are to the Soviet and Armenian Academies of Science and to Professor Ambartsumian for helping to make this important meeting possible. Doctor Morrison has some closing remarks.

MORRISON: I shall try very rapidly to make a final calculation. If you are having a meeting to discuss CETI, you should have a physical factor—blue skies, blue lake, snowy peaks; a historical factor—ancient writing and modern hydroelectric plants; and, a scientific factor—it is fine to work in a place where the spirit of research is alive. Finally, it is indispensable to have a human factor—warmth and welcome.

Obviously, the product of these probabilities should be very small. Yet all these factors were present at the Byurakan Observatory in the Armenian Republic. It is a very good omen. On behalf of the American Organizing Committee and I am sure many others, we say thank you very much. We hope someday to come back.

AMBARTSUMIAN: Thank you very much. We are deeply moved by your words and we hope that all of you will come here again. And now I would like to wish you a happy journey home. Bon voyage.

CONFERENCE RESOLUTIONS: FIRST SOVIET-AMERICAN CONFERENCE ON COMMUNICATION WITH EXTRATERRESTRIAL INTELLIGENCE (CETI)

The first international conference on the problem of extrater-restial civilizations, and contact with them, was held September 5–11, 1971, at the Byurakan Astrophysical Observatory of the Armenian Academy of Sciences, U.S.S.R. The Conference was a gathering of qualified scientists working in a variety of fields —astronomy, physics, radiophysics, computer science and technology, chemistry, biology, linguistics, archaeology, an-thropology, sociology, and history—and included many dis-tinguished scientists. The conference was jointly organized by the U.S. National Academy of Sciences (with assistance from the U.S. National Science Foundation) and the U.S.S.R. Academy of Sciences. Scientists from several other nations par-ticipated.

Many aspects of the problem of extraterrestrial civilizations were discussed in detail in the ten sessions of the Conference. Particular attention was devoted to the following questions: the plurality of planetary systems in the universe, the origin of life on Earth, the possibility of life arising on cosmic bodies, the origin and evolution of intelligence, the origin and development of technological civilizations, problems in searching for intelligent signals or for evidence of astroengineering activities, and the problems and possible consequences of establishing contact with extraterrestrial civilizations.

Conference participants differed on many details of these questions, but were agreed that the promise of contact with such extraterrestrial civilizations is sufficiently high to justify initiating a variety of well-formulated search programs; they also agreed that present technology may be capable of establishing contact with such civilizations. Some preliminary radio-astronomical searches have been performed both in the U.S. and in the U.S.S.R.

The Conference participants reached the following conclusions:

1. The striking discoveries of recent years in the fields of astronomy, biology, computer science, and radiophysics have transferred some of the problems of extraterrestrial civilizations and their detection from the realm of speculation to a new realm of experiment and observation. For the first time in human history it has become possible to make serious and detailed experimental investigations of this fundamental and important problem.

2. This problem may prove to be of profound significance for the future development of mankind. If extraterrestrial civilizations are ever discovered the effect on human scientific and technological capabilities will be immense, and the discovery can positively influence the whole future of man. The practical and philosophical significance of a successful contact with an extraterrestrial civilization would be so enormous as to justify the expenditure of substantial efforts. The consequences of such a discovery would greatly add to the total of human knowledge.

3. The technological and scientific resources of our planet are already large enough to permit us to begin investigations directed

towards the search for extraterrestrial intelligence. As a rule, such studies should provide important scientific results even when specific searches for extraterrestrial intelligence do not succeed. At present these investigations can be carried out effectively in the various countries by their own scientific institutions. Even at this early stage, however, it would be useful to discuss and coordinate specific programs of research and to exchange scientific information. In the future it would be desirable to combine the efforts of investigators in various countries to achieve the experimental and observational objectives. It seems to us appropriate that the search for extraterrestrial intelligence should be made by representatives of the whole of mankind.

4. Various modes of search for extraterrestrial intelligence were discussed in detail at the Conference. The realization of the most elaborate of these proposals would require considerable time and effort and an expenditure of funds comparable to the funds devoted to space and nuclear research. Useful searches can, however, also be initiated at a very modest scale.

5. The Conference participants consider highly valuable present and forthcoming space-vehicle experiments directed toward searching for life on the other planets of our solar system. They recommend the continuation and strengthening of work in such areas as prebiological organic chemistry, searches for extrasolar planetary systems, and evolutionary biology, which bear sharply on the problem.

6. The Conference recommends the initiation of specific new investigations directed towards modes of search for signals. A list of some possible investigations is appended.

7. To coordinate national programs of research and to promote progress in this field, the Conference suggests the establishment by appropriate means of an international working group. For the time being, the following interim working group is pro-

posed: F. Drake, U.S.A.; N. S. Kardashev, U.S.S.R.; P. Morrison, U.S.A.; B. Oliver, U.S.A.; R. Pešek, Czechoslovakia; C. Sagan, U.S.A.; I. S. Shklovsky, U.S.S.R.; G. M. Tovmasyan, U.S.S.R.; and V. S. Troitsky, U.S.S.R.

8. The Conference participants urge the full and open publication of research results on these problems, and as a step in this direction plan simultaneous publication in Russian and in English of the Proceedings of the present Conference.

9. The interim working group is instructed to consider, when necessary, convening broadly based or more specialized meetings of scientists working on CETI.

10. The Conference participants express their warm appreciation for the splendid hospitality extended to them by the Armenian Academy of Sciences.

[Signed]

The Organizing Committees of the U.S. and U.S.S.R. delegations, for the Conference participants

List of Possible Research Directions
It would be useful to concentrate efforts in two directions, both of which seem promising:
I. Searches for civilizations at a technical level comparable with our own.
II. Searches for civilizations at a technological level greatly surpassing our own.
A wide circle of specialists, from astrophysicists to historians, should participate in the planning of this research.

Accordingly, we recommend:

1. A search for signals and for evidence of astroengineering activities in the radiation of a few hundred chosen nearby stars

and of a limited number of other selected objects, covering the wavelength range from visible to decimeter waves, using the largest existing astronomical instruments.

2. A search for signals from powerful sources within galaxies of the local group, including searches for strong impulsive signals.

3. Exploration of the region of minimum noise in the submillimeter band, in order to determine its suitability for observing extraterrestrial civilizations.

The following studies are desirable:

4. The design, among others, of powerful new astronomical instruments with roughly the following parameters:

(a) A decimeter wave radio telescope with effective area $\gtrsim 1$ square kilometer.

(b) A millimeter wave telescope with effective area $> 10^4$ square meters

(c) A submillimeter wave telescope with effective area $\gtrsim 10^3$ square meters.

(d) An infrared telescope with effective area $\gtrsim 10^2$ square meters.

All of the instruments described above have the capability of providing important data in subjects quite separate from CETI.

5. The definition of a system for keeping the entire sky under constant surveillance, which could lead to a search of wider scope than those listed under 1 and 2.

APPENDIX A

NATURE OF PROBABILITY STATEMENTS IN DISCUSSIONS OF THE PREVALENCE OF EXTRATERRESTRIAL INTELLIGENT LIFE

TERRENCE FINE
School of Electrical Engineering
Cornell University, Ithaca, New York

Introduction

Attempts to make meaningful probability statements about the prevalence of extraterrestrial intelligent life (ETIL) have experienced significant criticism and encountered considerable obstacles. It is our view that progress on this question will require an explicit examination of the various types of probability statements and concepts so as to reject the patently unsuitable ones and select reasonable and tractable forms. A general discussion of different theories of probability is available in Fine (1972). Our hasty overview of the nature of probability will suggest that at present the most realistic type of probability statement for questions concerning ETIL has a comparative form and a subjective interpretive basis.

Types of Probability Statements

Probability statements can be distinguished with respect to their forms and their interpretive bases.

Forms Three basic forms are

Quantitative—assigns a number in [0, 1] to each event or statement in a suitable collection (σ-algebra), e.g. "P(there is ETIL) = 0.6";

Comparative—determines which of a pair of events is the more probable but not by how much; e.g., "ETIL in our Galaxy is more probable than in Andromeda";

Modal—merely identifies the probable events; e.g. "ETIL in our Galaxy is probable."

While the quantitative form is the most common in scientific practice, the more modest comparative and modal forms tend to be more realistic descriptions of uncertainty. Despite possible appearances, the modal and comparative forms are not in general reducible to quantitative terms; while they are weaker forms, they are more generally applicable. It seems reasonable that statements about ETIL might be better formulated in comparative terms, which can still be fruitfully used in making decisions concerning, say, desirable directions of research, than in the more specific, and thus more elusive, quantitative form. Further discussion of comparative and modal forms of probability statements is available in Carnap (1962) and in Fine (1972).

Interpretive Bases The interpretive basis of a concept of probability indicates how it is to be measured, assessed, or estimated and how it can be correctly used or applied. The three root interpretations, each having several variants, are frequency of occurrence of events in repeated, unrelated trials; logical degree of confirmation of an hypothesis by evidence; and subjective, introspective statements of belief.

Frequency Interpretation

The frequency interpretation is the most widely espoused in the sciences; e.g., having found ETIL in p percent of the many systems of a given class that we have examined, we "estimate" that the probability that the next system of this class to be examined will possess ETIL is "approximately" p. The frequency approach tries to identify and then directly extrapolate trends in patterns of occurrence. It is generally thought to yield an objective and empirical concept of probability and is therefore suitable for science. However, sensitive physical scientists and many philosophers have held that the frequency approach is ambiguous and its objectivity is illusory. To quote Feynman (1963):

"Probability depends, therefore, on our knowledge and on our ability to make estimates. In effect, on our common sense!"

[6-2]

"It is probably better to realize that the probability concept is in a sense subjective, . . ."   [6-7]

How do we recognize whether nonidentical systems are of the same class for purposes of extrapolation? Which of the many patterns or trends should we extrapolate? What guarantees the persistence of a particular trend? In practice, estimators that provide the working definitions of frequency-based probability have large subjective and nonempirical components involving personal choices of statistical procedures and selection and censoring of data. Finally, it is evident that the frequency conception has no bearing on the issues currently raised by the question of ETIL; the requisite data, whatever it might really be, is clearly absent in this case.

Logical Interpretation
The logical interpretation of probability attempts to supply, through an analysis of inductive reasoning, an objective, logical (analytic) procedure for the determination of the degree to

which a set of data statements supports or confirms an hypothesis or theory. The muddled classical or Laplacian theory of probability (based upon equally likely cases adduced from a balance of evidence) is probably the best-known informal theory of quantitative logical probability. The most fully developed theories of logical probability are due to Carnap (1962) and to Solomonoff (1964) and are as yet too underdeveloped for significant application. While the notion of an objective logical degree of support relating data and theory is attractive and relevant to the question of ETIL, there is no present prospect that this approach will be developed to the point where it is applicable to such complex scientific questions. Nor can much more be expected from the present primitive modal and comparative logical theories.

Subjective Interpretation
The subjective or personalistic interpretation of probability championed by Savage (1954), Pratt, Raiffa, and Schlaifer (1965), and others maintains that probability statements are derived through a largely unassisted process of introspection and are then applied to the selection of optimal decisions or acts, such as the allocation of research resources. The subjective view boldly admits of the subjective elements in most other concepts of probability and encourages the holder to fully use his informal judgment, beliefs, experience in arriving at probability estimates whose objective is decision making and the interpersonal communication of individual judgment and not, say, assessing the "truth" of propositions. While subjective probability statements are indeed personal, they are not arbitrary. There are reasonable axioms of internal consistency between assessments and constraints that force the user to learn from experience in a reasonably explicit way. While it is not possible to criticize any single

subjective probability assessment it is possible to criticize a collection of such assessments.

The subjective approach has been widely discussed and applied in management decision making (Schlaifer, 1969) and in reliability analysis, although in few other cases. While it has evident limitations, so do all of the other approaches to probability. We would judge that the concept of subjective probability is at present the only basis upon which probability statements can be made about ETIL.

References

Carnap, R. (1962). *Logical Foundations of Probability,* 2nd ed. Chicago: University of Chicago Press.

Feynman, R., et al. (1963). *The Feynman Lectures on Physics,* v. I. Reading, Mass.: Addison-Wesley.

Fine, T. (1972). *Theories of Probability: An Examination of Foundations.* New York: Academic Press. 1972.

Pratt, J., H. Raiffa, R. Schlaifer (1965). *Introduction to Statistical Decision Theory.* New York: McGraw-Hill.

Savage, L. (1954). *The Foundations of Statistics.* New York: Wiley.

Schlaifer, R. (1969). *Analysis of Decisions Under Uncertainty.* New York: McGraw-Hill.

Solomonoff, R. (1964). A Formal Theory of Inductive Inference, Part I. *Information and Control* 7: 1–22.

APPENDIX B

ADDED COMMENTS ON
"THE NONPREVALENCE OF HUMANOIDS" *

G. G. SIMPSON
Department of Vertebrate Palaeontology
Harvard University, Cambridge, Massachusetts, and
University of Arizona, Tucson, Arizona

The question now posed about "the likelihood of the origin of intelligence, not necessarily even remotely human" may make my discussion of "humanoids" seem irrelevant. In fact the question is worthy of a "science" without known subject matter. Intelligence not even remotely human is like sight not even remotely involving vision, or, better, communication between beings with no possible means of understanding each other. My discussion of humanoids postulates enough similarity to permit communication, and no more. If communication is impossible, the quest for signs of life outside the solar system is now impossible for that if for no other reason.

Knowledge of Venus and Mars has increased considerably in the last eight years. It confirms the virtual impossibility of carbon-based life on Venus and greatly reduces the chances of such life on Mars. The chances of intelligent life (even *remotely* hu-

* Cf. G. G. Simpson, The Nonprevalence of Humanoids. *Science* 143 (1964): 769.

man) on either planet, or any other body in our solar system, are evidently as near nil as possible.

Evidence for spontaneous and deterministic origin of prebiological organic molecules continues to increase. Evidence on details of the really crucial step from that to cellular (true) life continues to be virtually lacking, even though we know that has occurred once.

There is now good evidence that cellular organisms existed on earth not merely two but well over three billion years ago. That *decreases* the probability of a parallel origin of intelligence elsewhere.

A number of other biologists have independently calculated that such a probability is almost vanishingly small (e.g., H. Blum, *Nature* 206:131; Slater, Proc. VIII Astronaut. Cong.: 395; and others). Astronomers, physicists, and chemists continue to be the main supporters of the idea, in their ignorance of biology, and even some highly respectable authorities in those fields ridicule it (e.g., D. Menzel, *Grad. Journal* 7:195).

If the universe is infinite, the numbers of stars and possible planets is also infinite and the wildly varying estimates of those numbers are ridiculous. The estimates have meaning only if they refer to the number we might communicate with. These estimates vary enormously, but there is good reason to doubt the higher claims of numbers of accessible planets (e.g., S. Kumar, An. New York Acad. Sci., 163:94). And it is still true that not even one earthlike planet outside our solar system has been observed objectively or is known to exist in fact. Exobiology is still a "science" without *any* data, therefore no science.

Some otherwise respectable scientists, like a greatly lamented colleague at the University of Arizona who recently committed suicide, have continued to believe that some UFO's have had extraterrestrial guidance. That continues to be a monument to gullibility. The "evidence" for witchcraft is much better. Cogent antireferences are legion; for just one example see W. Markowitz, *Science* 157:1274.

It is still true that no credible evidence of fossil organisms in meteorites has been advanced, but some further credible but not conclusive evidence of probably original organic compounds in meteorites has been obtained. A distinction must be made between "organic" and "biogenic." There is still no credible evidence that these compounds were biogenic, and that claim has not recently been reiterated.

After publication of my 1964 *Science* article I was accused of misrepresenting expenditures on exploration for extraterrestrial life. As anyone who actually reads that article can verify, I did not in fact say that *any* expenditure was being made for such exploration, although in fact a considerable amount was. What I did say was that the prospective discovery of extraterrestrial life was being advanced as a reason or excuse for space exploration in general, on which billions of dollars were being spent. That is easy enough to verify; see, for example, the NAS-NRC report on "Biology and the Exploration of Mars," C. S. Pittendrigh, Chairman. The billions have now been slowed down, but they are still billions, and "exobiology" is still part of the argument, whether or not the "exobiologists" feel they are getting their share.

A friendly reviewer (Philip Morrison) of the well-known book by Shklovskii and Sagan on *Intelligent Life in the Universe* wrote that "here is a body of literature whose ratio of results/papers is lower than any other." I suggest that the participants think of this as still another conference on the subject is convened.

APPENDIX C

ON THE DETECTIVITY OF ADVANCED GALACTIC CIVILIZATIONS *

CARL SAGAN
Laboratory for Planetary Studies
Cornell University, Ithaca, New York

Mankind now possesses the technological capability of communicating at radiofrequencies, over distances of many hundreds of light years, with technical civilizations no more advanced than we. But before a program is initiated to search systematically for such signals it is important to demonstrate at least a modest probability that one technical civilization exists within such a range. The possibility that much more advanced civilizations exist—societies which can be detected over much larger distances—will be discussed presently. The pitfalls in placing numerical values on the component probabilities of N, the number of extant technical civilizations in the Galaxy, are numerous and treacherous; nevertheless, there does seem to be a limiting factor whose significance has not always been appreciated.

While much more sophisticated formulations are now available (see, e.g., Kreifeldt, 1971), the first algebraic expression for N, due in its original formulation to F. D. Drake, will serve our purpose:

* Reprinted from *Icarus* 19 (1973): 350–352.

$$N = RL. \tag{1}$$

Here R is the rate of emergence of communicative technical civilizations in the Galaxy and is a function of the rate of star formation, the fraction of stars which have planets, the number of planets per star which are ecologically suitable for the origin of life, the fraction of such planets on which the origin of life actually occurs, and the fraction of such planets on which intelligence and eventually technological civilizations actually emerge (see, e.g., Shklovskii and Sagan, 1966; henceforth, reference S^2); L is the mean lifetime of such civilizations, and is strongly biased toward the small fraction of technical civilizations which achieve very long lifetimes—lifetimes measured on the geological or stellar evolutionary time scales (Ref. S^2). But such civilizations will be inconceivably in advance of our own. We have only to consider the changes in mankind in the last 10^4 years and the potential difficulties which our Pleistocene ancestors would have in accommodating to our present society to realize what an unfathomable cultural gap 10^8 to 10^{10} years represents, even with a tiny rate of intellectual advance. Such societies will have discovered laws of nature and invented technologies whose applications will appear to us indistinguishable from magic. There is a serious question about whether such societies are concerned with communicating with us, any more than we are concerned with communicating with our protozoan or bacterial forebears. We may study microorganisms, but we do not usually communicate with them. I therefore raise the possibility that a horizon in communications interest exists in the evolution of technological societies, and that a civilization very much more advanced than we will be engaged in a busy communications traffic with its peers; but not with us, and not via technologies accessible to us. We may be like the inhabitants of the valleys of New Guinea who may communicate by runner or drum, but who are ignorant of the vast international radio and cable traffic passing over, around and through them.

A convenient subdivision of galactic technological societies has been provided by Kardashev (1962). He distinguishes Type I, Type II, and Type III civilizations. The first is able to engage something like the present power output of the planet Earth for interstellar discourse; the second the power output of a sun; and the third the power output of a galaxy. By definition, Type I civilizations are capable of restructuring planets, Type II civilizations of restructuring solar systems, and Type III civilizations of restructuring galaxies. I believe that a civilization of approximately Type II has, with an exception to be described later, reached our communications horizon. For computational convenience, I also assume that a civilization which has emerged to Type II technologies has also successfully passed through the critical period of probable technological self-destruction—the period in which terrestrial civilization is now immersed.

These ideas can now be restated as follows: let f_g be the fraction of technical civilizations which survive for geological or stellar evolutionary time scales L_g, and let L_d be the mean time to self-destruction of those Type I civilizations which do not achieve Type II technologies. Then,

$$L \sim (1 - f_g)L_d + f_g L_g. \tag{2}$$

Accordingly, the total number of extant civilizations in the Galaxy,

$$N \sim R[(1 - f_g)L_d + f_g L_g], \tag{3}$$

is different from the number of civilizations within our communications horizon,

$$N_c \sim N_I \sim R[(1 - f_g)L_d]. \tag{4}$$

The ratio of these lifetimes

$$N_c/N \sim [1 + f_g(L_g/L_d)]^{-1}, \qquad \text{for } f_g \ll 1 \qquad (5)$$

$$\sim (L_d/f_g L_g), \qquad \text{for } f_g L_g \gg L_d. \qquad (6)$$

Equations (5) and (6) are independent of R. Of the civilizations within our communications horizon only

$$N_c' \sim R f_g (1 - f_g) L_d \qquad (7)$$

are destined to have lifetimes $\gg L_d$.

We now specialize to some illustrative numerical cases. I emphasize that values differing by several orders of magnitude from the ones I choose are certainly conceivable and may even be probable. We adopt (Ref. S[2]) $L_g \sim 10^9$ years, $f_g \sim 10^{-2}$, and $R \sim 10^{-1}$ per year. I further assume $L_d \sim 10^8$ years. From events of the past few decades a case can be made for L_d to be 1 to 2 orders of magnitude smaller; the resulting conclusions will be correspondingly more pessimistic. Independent of the choice of L_d, as long as $L_d \ll L_g$, we find $L \sim 10^7$ years, and $N \sim 10^6$ galactic civilizations (Ref. S[2]). Assuming such civilizations are randomly distributed, the mean distance to the nearest is a few hundred light years, and searches for such civilizations, using existing technology, would seem to be in order. However, if we count only those civilizations within our communications horizon, we find, with the same choice of numbers,

$$N_c/N \sim 10^{-4}$$

and

$$N_c \sim 100.$$

In this case the distance to the nearest communicative civilization is $\sim 10^4$ light years—well beyond our present capability, assuming that our communicant is at approximately the same technological level. And of these 100 societies only $N_c' \sim 1$ is likely to avoid self-destruction.

Almost all of these 100 civilizations of Type I or younger must have technologies significantly in advance of our own, and it may very well be possible to make contact with them. But the prospects are very much dimmer than in the case of 10^6 communicative galactic civilizations. The situation can be improved somewhat by taking $L_d <$ the interval to the communications horizon, rather than equal to it as we have assumed here; but we have been optimistic in our choice of L_d and I find it difficult to imagine that many civilizations $> 10^3$ years in our technological future would be anxious to communicate with us.

The situation seems to be that Type II or more advanced civilizations may be, in terms of contemporary terrestrial communications technology, at small distances from us—but, in the same terms, noncommunicative; whereas Type I civilizations may be communicating—but tend to be too far away for us to detect. The operational consequence is that the detection of civilizations of Type I or younger is more difficult than has generally been assumed, and that such an enterprise will require much more elaborate radio systems—for example, very large phased arrays—than currently exist, and very long observing times to search through the $\sim 10^9$ stars which must be winnowed to find one such civilization.

On the other hand somewhat more serious attention must be given to the question of Type II and Type III civilizations, the level where, according to the previous argument, most of the technical societies in the universe are. A Type II civilization can communicate with the earth from our nearest galactic neighbors; a Type III civilization can communicate across the known universe—and this employing only laws of nature which we now understand. If only a tiny fraction of such civilizations are interested in antique communications modes they will dominate the interstellar communications traffic now accessible on Earth. The best policy might therefore be to search with existing technology for Type II or Type III civilizations among the nearer galaxies, rather than Type I or younger civilizations among the nearer stars.

References

Kardashev, N. S. (1964). Transmission of Information by Extraterrestrial Civilizations. *Astrononicheskii Zh.* 41: 282 [English translation in *Soviet Astronomy—A. J.* 8: 217].

Kreifeldt, J. G. (1971). A Formulation for the Number of Communicative Civilizations in the Galaxy. *Icarus* 14: 419–430.

Shklovskii, I. S., and Carl Sagan (1966) (Ref. S²). *Intelligent Life in the Universe.* San Francisco: Holden-Day.

APPENDIX D

THE WORLD, THE FLESH, AND THE DEVIL

FREEMAN J. DYSON
Institute for Advanced Study
Princeton, New Jersey

I. Bernal's Book
The World, the Flesh and the Devil; an Enquiry into the Future
of the Three Enemies of the Rational Soul, is the full title of
Bernal's first book which he wrote at the age of 28. Forty years
later he said in a foreword to the second edition, "This short
book was the first I ever wrote. I have a great attachment to it
because it contains many of the seeds of ideas which I have been
elaborating throughout my scientific life. It still seems to me to
have validity in its own right." It must have been a consolation
to Bernal, crippled and incapacitated in the last years of his life,
to know that this work of his spring-time was again being bought
and read by a new generation of young readers.

Bernal's book begins with these words: "There are two futures,
the future of desire and the future of fate, and man's reason has
never learnt to separate them." I do not know of any finer open-

Third J. D. Bernal Lecture delivered at Birkbeck College, London, 16th
May 1972. Printed for private circulation by Birkbeck College, 1972.
Reprinted by kind permission of the Master of Birkbeck College.

ing sentence of a work of literature in English. Bernal's modest claim that his book "still seems to have validity in its own right" holds good in 1972 as it did in 1968. Enormous changes have occurred since he wrote the book in 1929, both in science and in human affairs. It would be miraculous if nothing in it had become dated or superseded by the events of the last forty years. But astonishingly little of it has proved to be wrong or irrelevant to our present concerns.

I decided that the best way I can do honor to Bernal in this lecture is to use his book as a point of departure for my own speculations about the future of mankind. I shall not expound or criticize the book in detail. I hope that much of what I shall say will be fresh and will go in some directions beyond Bernal's horizons. But it will be obvious to those of my audience who have read Bernal that my ideas are deeply influenced by him. To those who have not read Bernal I hope that I may provide a stimulus to do so.

Bernal saw the future as a struggle of the rational side of man's nature against three enemies. The first enemy he called the World, meaning scarcity of material goods, inadequate land, harsh climate, desert, swamp, and other physical obstacles which condemn the majority of mankind to lives of poverty. The second enemy he called the Flesh, meaning the defects in man's physiology that expose him to disease, cloud the clarity of his mind, and finally destroy him by senile deterioration. The third enemy he called the Devil, meaning the irrational forces in man's psychological nature that distort his perceptions and lead him astray with crazy hopes and fears, overriding the feeble voice of reason. Bernal had faith that the rational soul of man would ultimately prevail over these enemies. But he did not foresee cheap or easy victories. In each of the three struggles, he saw hope of defeating the enemy only if mankind is prepared to adopt extremely radical measures.

Briefly summarized, the radical measures which Bernal prescribed were the following. To defeat the World, the greater part of the human species will leave this planet and go to live in innumerable freely floating colonies scattered through outer space. To defeat the Flesh, humans will learn to replace failing organs with artificial substitutes until we become an intimate symbiosis of brain and machine. To defeat the Devil, we shall first reorganize society along scientific lines, and later learn to exercise conscious intellectual control over our moods and emotional drives, intervening directly in the affective functions of our brains with technical means yet to be discovered. This summary is a crude oversimplification of Bernal's discussion. He did not imagine that these remedies would provide a final solution to the problems of humanity. He well knew that every change in the human situation will create new problems and new enemies of the rational soul. He stopped where he stopped because he could not see any farther. His chapter on "The Flesh" ends with the words: "That may be an end or a beginning, but from here it is out of sight."

How much that was out of sight to Bernal in 1929 can we see from the vantage-point of 1972? The first and most obvious difference between 1929 and 1972 is that we have now a highly vocal and well-organized opposition to the further growth of the part that technology plays in human affairs. The social prophets of today look upon technology as a destructive rather than a liberating force. In 1972 it is highly unfashionable to believe as Bernal did that the colonization of space, the perfection of artificial organs and the mastery of brain physiology are the keys to man's future. Young people in tune with the mood of the times regard space as irrelevant, and they consider ecology to be the only branch of science that is ethically respectable. However, it would be wrong to imagine that Bernal's ideas were more in line with popular views in 1929 than they are in 1972. Bernal was never a man to swim with the tide. Technology was unpopular in

1929 because it was associated in people's minds with the gas warfare of the first World War, just as now it is unpopular by association with Hiroshima and the defoliation of Vietnam. In 1929 the dislike of technology was less noisy than today but no less real. Bernal understood that his proposals for the remaking of man and society flew in the teeth of deeply entrenched human instincts. He did not on that account weaken or compromise his statement. He believed that a rational soul would ultimately come to accept his vision of the future as reasonable, and that for him was enough. He foresaw that mankind might split into two species, one following the technological path which he described, the other holding on as best it could to the ancient folkways of natural living. And he recognized that the dispersion of mankind into the vastness of space is precisely what is required for such a split of the species to occur without intolerable strife and social disruption. The wider perspective which we have gained between 1929 and 1972 concerning the harmful effects of technology affects only the details and not the core of Bernal's argument.

Another conspicuous difference between 1929 and 1972 is that men have now visited the moon. Surprisingly, this fact makes little difference to the plausibility of Bernal's vision of the future. Bernal in 1929 foresaw cheap and massive emigration of human beings from the earth. He did not know in detail how it should be done. We still do not know how it should be done. Certainly it will not be done by using the techniques that took men to the moon in 1969. We know that in principle the cost in energy or fuel of transporting people from Earth into space need be no greater than the cost of transporting them from New York to London. To translate this "in principle" into reality will require two things: first a great advance in the engineering of hypersonic aircraft, and second the growth of a traffic massive enough to permit large economies of scale. It is likely that the Apollo vehicle bears the same relation to the cheap mass-transportation space-vehicle of the future as the majestic airship of the 1930s bears to the Boeing 747 of today. The airship R101 was absurdly large, beautiful, expensive, and fragile, just like the Apollo Saturn 5. If this analogy is sound, and I believe it is, we shall have transportation into space at a reasonable price within about fifty years from now. But my grounds for believing this are not essentially firmer than Bernal's were for believing it in 1929.

II. The Double Helix

The decisive change that has enabled us to see farther in 1972 than we could in 1929 is the advent of molecular biology. Bernal recognized this in the 1968 foreword to his book, where he speaks of the double helix as "the greatest and most comprehensive idea in all science." We now understand the basic principles by which living cells organize and reproduce themselves. Many mysteries remain, but it is inevitable that we shall understand the chemical processes of life in full detail, including the processes of development and differentiation of higher organisms, within the next century. I consider it also inevitable and desirable that we shall learn to exploit these processes for our own purposes. The next century will see a completely new technology growing out of the mastery of the principles of biology in the same way as our existing technology grew out of a mastery of the principles of physics.

The new biological technology may grow in three distinct directions. Probably all three will be followed and will prove fruitful for particular purposes. The first direction is the one that has been chiefly discussed by biologists who feel responsibility for the human consequences of their work; they call it "genetic surgery." The idea is that we shall be able to read the base-sequence of the DNA in a human sperm or egg-cell, run the sequence through a computer which will identify deleterious genes or mutations, and then by micromanipulation patch harmless genes into the sequence to replace the bad ones. It might also be possible to add to the DNA genes conferring various desired characteristics to the resulting individual. This technology will be difficult and dangerous, and its use will raise severe ethical problems. Jacques Monod in his recent book *Chance and Necessity* sweeps all thought of it aside with his customary dogmatic certitude. "There are," he says, "occasional promises of remedies expected from the current advances in molecular genetics. This illusion, spread about by a few superficial minds, had better be disposed of." Although I have a great respect for Jacques Monod, I still dare to brave his scorn by stating my belief that genetic surgery has an important part to play in man's future. But I share the prevailing view of biologists that we must be exceedingly careful in interfering with the human genetic material. The interactions between the thousands of genes in a human cell are so exquisitely complicated that a computer program labeling genes "good" or "bad" will be adequate to deal only with the grossest sort of defect. There are strong arguments for declaring a moratorium on genetic surgery for the next hundred years, or until we understand human genetics vastly better than we do now.

Leaving aside genetic surgery applied to humans, I foresee that the coming century will place in our hands two other forms of biological technology which are less dangerous but still revolutionary enough to transform the conditions of our existence. I count these new technologies as powerful allies in the attack on Bernal's three enemies. I give them the names "biological engineering" and "self-reproducing machinery." Biological engineering means the artificial synthesis of living organisms designed to fulfill human purposes. Self-reproducing machinery means the imitation of the function and reproduction of a living organism with nonliving materials, a computer program imitating the function of DNA and a miniature factory imitating the functions of protein molecules. After we have attained a complete understanding of the principles of organization and development of a simple multicellular organism, both of these avenues of technological exploitation should be open to us.

III. Biological Engineering

I would expect the earliest and least controversial triumphs of biological engineering to be extensions of the art of industrial fermentation. When we are able to produce microorganisms equipped with enzyme systems tailored to our own design, we can use such organisms to perform chemical operations with far greater delicacy and economy than present industrial practices allow. For example, oil refineries would contain a variety of bugs designed to metabolize crude petroleum into the precise hydrocarbon stereo-isomers which are needed for various purposes. One tank would contain the n-octane bug, another the benzene bug, and so on. All the bugs would contain enzymes metabolizing sulphur into elemental form, so that pollution of the atmosphere by sulphurous gases would be completely controlled. The management and operation of such fermentation tanks on a vast scale would not be easy, but the economic and social rewards are so great that I am confident we shall learn how to do it. After we have mastered the biological oil refinery, more important applications of the same principles will follow. We shall have factories producing specific foodstuffs biologically from cheap raw materials, and sewage-treatment plants converting our wastes efficiently into usable solids and pure water. To perform these operations we shall need an armamentarium of many species of microorganisms trained to ingest and excrete the appropriate chemicals. And we shall design into the metabolism of these organisms the essential property of self-liquidation, so that when deprived of food they disappear by cannibalizing one another.

They will not, like the bacteria that feed upon our sewage in today's technology, leave their rotting carcasses behind to make a sludge only slightly less noxious than the mess that they have eaten.

If these expectations are fulfilled, the advent of biological technology will help enormously in the establishment of patterns of industrial development with which human beings can live in health and comfort. Oil refineries need not stink. Rivers need not be sewers. However, there are many environmental problems which the use of artificial organisms in enclosed tanks will not touch. For example, the fouling of the environment by mining and by abandoned automobiles will not be reduced by building cleaner factories. The second step in biological engineering, after the enclosed biological factory, is to let artificial organisms loose into the environment. This is admittedly a more dangerous and problematical step than the first. The second step should be taken only when we have a deep understanding of its ecological consequences. Nevertheless the advantages which artificial organisms offer in the environmental domain are so great that we are unlikely to forego their use forever.

The two great functions which artificial organisms promise to perform for us when let loose upon the earth are mining and scavenging. The beauty of a natural landscape undisturbed by man is largely due to the fact that the natural organisms in a balanced ecology are excellent miners and scavengers. Mining is mostly done by plants and microorganisms extracting minerals from water, air, and soil. For example, it has been recently discovered that organisms in the ground mine ammonia and carbon monoxide from air with high efficiency. To the scavengers we owe the fact that a natural forest is not piled as high with dead birds as one of our junk yards with dead cars. Many of the worst offenses of human beings against natural beauty are due to our incompetence in mining and scavenging. Natural organisms know how to mine and scavenge effectively in a natural environment. In a man-made environment, neither they nor we know how to do it. But there is no reason why we should not be able to design artificial organisms that are adaptable enough to collect our raw materials and to dispose of our refuse in an environment that is a careful mixture of natural and artificial.

A simple example of a problem that an artificial organism could solve is the eutrophication of lakes. At present many lakes are being ruined by excessive growth of algae feeding on high levels of nitrogen or phosphorus in the water. The damage could be stopped by an organism that would convert nitrogen to molecular form or phosphorus to an insoluble solid. Alternatively and preferably, an organism could be designed to divert the nitrogen and phosphorus into a food chain culminating in some species of palatable fish. To control and harvest the mineral resources of the lake in this way will in the long run be more feasible than to maintain artificially a state of "natural" barrenness.

The artificial mining organisms would not operate in the style of human miners. Many of them would be designed to mine the ocean. For example, oysters might extract gold from seawater and secrete golden pearls. A less poetic but more practical possibility is the artificial coral that builds a reef rich in copper or magnesium. Other mining organisms would burrow like earthworms into mud and clay, concentrating in their bodies the ores of aluminum or tin or iron, and excreting the ores in some manner convenient for human harvesting. Almost every raw material necessary for our existence can be mined from ocean, air or clay, without digging deep into the earth. Where conventional mining is necessary, artificial organisms can still be useful for digesting and purifying the ore.

Not much imagination is needed to foresee the effectiveness of artificial organisms as scavengers. A suitable microorganism could convert the dangerous organic mercury in our rivers and lakes to a harmless insoluble solid. We could make good use of an organism with a consuming appetite for polyvinyl chloride and similar plastic materials which now litter beaches all over the earth. Conceivably we may produce an animal specifically designed for chewing up dead automobiles. But one may hope that the automobile in its present form will become extinct before it needs to be incorporated into an artificial foodchain. A more serious and permanent role for scavenging organisms is the removal of trace quantities of radioactivity from the environment. The three most hazardous radioactive elements produced in fission reactors are strontium, cesium, and plutonium. These elements have long half-lives and will inevitably be released in small quantities so long as mankind uses nuclear fission as an energy source. The long-term hazard of nuclear energy would be notably reduced if we had organisms designed to gobble up these three elements from water or soil and to convert them into indigestible form. Fortunately, none of these three elements is essential to our body chemistry, and it therefore does us no harm if they are made indigestible.

IV. Big Trees

I have spoken about the two first steps of biological engineering. The first will transform our industry and the second will transform our earth-bound ecology. It is now time to speak of the third step, which is the colonization of space. I believe in fact that biological engineering is the essential tool which will make Bernal's dream of the expansion of mankind in space a practical possibility.

First I have to clear away a few popular misconceptions about space as a habitat. It is generally considered that planets are important. Except for Earth, they are not. Mars is waterless, and the others are for various reasons basically inhospitable to man. It is generally considered that beyond the sun's family of planets there is absolute emptiness extending for light years until you come to another star. In fact it is likely that the space around the solar system is populated by huge numbers of comets, small worlds a few miles in diameter, rich in water and the other chemicals essential to life. We see one of these comets only when it happens to suffer a random perturbation of its orbit which sends it plunging close to the sun. It seems that roughly one comet per year is captured into the region near the sun, where it eventually evaporates and disintegrates. If we assume that the supply of distant comets is sufficient to sustain this process over the thousands of millions of years that the solar system has existed, then the total population of comets loosely attached to the sun must be numbered in the thousands of millions. The combined surface area of these comets is then a thousand or ten thousand times that of Earth. I conclude from these facts that comets, not planets, are the major potential habitat of life in space. If it were true that other stars have as many comets as the sun, it then would follow that comets pervade our entire Galaxy. We have no evidence either supporting or contradicting this hypothesis. If true, it implies that our Galaxy is a much friendlier place for interstellar travelers than it is popularly supposed to be. The average distance between habitable oases in the desert of space is not measured in light years, but is of the order of a light day or less.

I propose to you then an optimistic view of the Galaxy as an abode of life. Countless millions of comets are out there, amply supplied with water, carbon, and nitrogen, the basic constituents of living cells. We see when they fall close to the sun that they contain all the common elements necessary to our existence. They lack only two essential requirements for human settlement, namely warmth and air. And now biological engineering will come to our rescue. We shall learn to grow trees on comets.

To make a tree grow in airless space by the light of a distant sun is basically a problem of redesigning the skin of its leaves. In every organism the skin is the crucial part which must be most delicately tailored to the demands of the environment. The skin of a leaf in space must satisfy four requirements. It must be opaque to far-ultraviolet radiation to protect the vital tissues from radiation damage. It must be impervious to water. It must transmit visible light to the organs of photosynthesis. It must have extremely low emissivity for far-infrared radiation, so that it can limit loss of heat and keep itself from freezing. A tree whose leaves possess such a skin should be able to take root and flourish upon any comet as near to the sun as the orbits of Jupiter and Saturn. Farther out than Saturn the sunlight is too feeble to keep a simple leaf warm, but trees can grow at far greater distances if they provide themselves with compound leaves. A compound leaf would consist of a photosynthetic part which is able to keep itself warm, together with a convex mirror part which itself remains cold but focuses concentrated sunlight upon the photosynthetic part. It should be possible to program the genetic instructions of a tree to produce such leaves and orient them correctly toward the sun. Many existing plants possess structures more complicated than this.

Once leaves can be made to function in space, the remaining parts of a tree—trunk, branches, and roots—do not present any great problems. The branches must not freeze, and therefore the bark must be a superior heat insulator. The roots will penetrate and gradually melt the frozen interior of the comet, and the tree will build its substance from the materials which the roots find there. The oxygen which the leaves manufacture must not be exhaled into space; instead it will be transported down to the roots and released into the regions where men will live and take their ease among the tree trunks. One question still remains. How high can a tree on a comet grow? The answer is

surprising. On any celestial body whose diameter is of the order
of ten miles or less, the force of gravity is so weak that a tree
can grow infinitely high. Ordinary wood is strong enough to lift
its own weight to an arbitrary distance from the center of gravity.
This means that from a comet of ten-mile diameter, trees can
grow out for hundreds of miles, collecting the energy of sunlight
from an area thousands of times as large as the area of the comet
itself. Seen from far away, the comet will look like a small po-
tato sprouting an immense growth of stems and foliage. When
man comes to live on the comets, he will find himself returning
to the arboreal existence of his ancestors.

We shall bring to the comets not only trees but a great variety
of other flora and fauna to create for ourselves an environment
as beautiful as ever existed on Earth. Perhaps we shall teach our
plants to make seeds which will sail out across the ocean of
space to propagate life upon comets still unvisited by man. Perhaps we shall start a wave of life which will spread from comet to
comet without end until we have achieved the greening of the
Galaxy. That may be an end or a beginning, as Bernal said, but
from here it is out of sight.

V. Self-Reproducing Machinery

In parallel with our exploitation of biological engineering, we
may achieve an equally profound industrial revolution by following the alternative route of self-reproducing machinery. Self-
reproducing machines are devices which have the multiplying
and self-organizing capabilities of living organisms but are built
of metal and computers instead of protoplasm and brains. It
was the mathematician John von Neumann who first demonstrated that self-reproducing machines are theoretically possible
and sketched the logical principles underlying their construction.
The basic components of a self-reproducing machine are precisely analogous to those of a living cell. The separation of

function between genetic material (DNA) and enzymatic machinery (protein) in a cell corresponds exactly to the separation between software (computer programs) and hardware (machine tools) in a self-reproducing machine.

I assume that in the next century, partly imitating the processes of life and partly improving on them, we shall learn to build self-reproducing machines programmed to multiply, differentiate, and coordinate their activities as skillfully as the cells of a higher organism such as a bird. After we have constructed a single egg machine and supplied it with the appropriate computer program, the egg and its progeny will grow into an industrial complex capable of performing economic tasks of arbitrary magnitude. It can build cities, plant gardens, construct electric power-generating facilities, launch space ships, or raise chickens. The overall programs and their execution will remain always under human control.

The effects of such a powerful and versatile technology on human affairs are not easy to foresee. Used unwisely, it offers a rapid road to ecological disaster. Used wisely, it offers a rapid alleviation of all the purely economic difficulties of mankind. It offers to rich and poor nations alike a rate of growth of economic resources so rapid that economic constraints will no longer be dominant in determining how people are to live. In some sense this technology will constitute a permanent solution of man's economic problems. Just as in the past, when economic problems cease to be pressing, we shall find no lack of fresh problems to take their place.

It may well happen that on Earth, for aesthetic or ecological reasons, the use of self-reproducing machines will be strictly limited and the methods of biological engineering will be used instead wherever this alternative is feasible. For example, self-reproducing machines could proliferate in the oceans and collect minerals for man's use, but we might prefer to have the same job done more quietly by corals and oysters. If economic needs were no longer paramount, we could afford a certain loss of efficiency for the sake of a harmonious environment. Self-reproducing machines may therefore play on Earth a subdued and self-effacing role.

The true realm of self-reproducing machinery will be in those regions of the solar system that are inhospitable to man. Machines built of iron, aluminum, and silicon have no need of water. They can flourish and proliferate on the moon or on Mars or among the asteroids, carrying out gigantic industrial projects at no risk to the earth's ecology. They will feed upon sunlight and rock, needing no other raw material for their construction. They will build in space the freely floating cities that Bernal imagined for human habitation. They will bring oceans of water from the satellites of the outer planets, where it is to be had in abundance, to the inner parts of the solar system where it is needed. Ultimately this water will make even the deserts of Mars bloom, and men will walk there under the open sky breathing air like the air of Earth.

Taking a long view into the future, I foresee a division of the solar system into two domains. The inner domain, where sunlight is abundant and water scarce, will be the domain of great machines and governmental enterprises. Here self-reproducing machines will be obedient slaves, and men will be organized in giant bureaucracies. Outside and beyond the sunlit zone will be the outer domain, where water is abundant and sunlight scarce. In the outer domain lie the comets where trees and men will live in smaller communities, isolated from each other by huge distances. Here men will find once again the wilderness that they have lost on Earth. Groups of people will be free to live as they please, independent of governmental authorities. Outside and away from the sun, they will be able to wander forever on the open frontier that this planet no longer possesses.

VI. Devils and Pilgrims

I have spoken much about how we may deal with the World and
the Flesh, and I have said nothing about how we may deal with
the Devil. Bernal also had difficulties with the Devil. He ad-
mitted in the 1968 foreword to his book that the chapter on the
Devil was the least satisfactory part of it. The Devil will always
find new varieties of human folly to frustrate our too rational
dreams.

Instead of pretending that I have an antidote to the Devil's
wiles, I will end this lecture with a discussion of the human factors that most obviously stand in the way of our achieving the
grand designs which I have been describing. When mankind is
faced with an opportunity to embark on any great undertaking,
there are always three main human factors that devilishly hamper
our efforts. The first is an inability to define or agree upon our
objectives. The second is an inability to raise sufficient funds.
The third is the fear of a disastrous failure. All three factors have
been conspicuously plaguing the United States space program
in recent years. It is a remarkable testimony to the vitality of
the program that these factors have still not succeeded in bring-
ing it to a halt. When we stand before the far greater enterprises
of biological technology and space colonization that lie in our
future, the same three factors will certainly rise again to con-
fuse and delay us.

I want now to demonstrate to you by a historical example how
these human factors may be overcome. I shall quote from
William Bradford, one of the Pilgrim Fathers, who wrote a book
called *Of Plimoth Plantation* describing the history of the first
English settlement in Massachusetts. Bradford was governor of
the Plymouth colony for 28 years. He began to write his history
ten years after the settlement. His purpose in writing it was, as
he said, "That their children may see with what difficulties their
fathers wrestled in going through these things in their first be-

ginnings. As also that some use may be made hereof in after times by others in such like weighty employments." Bradford's work remained unpublished for two hundred years, but he never doubted that he was writing for the ages.

Here is Bradford describing the problem of man's inability to agree upon objectives. The date is Spring 1620, the same year in which the Pilgrims were to sail.

But as in all businesses the acting part is most difficult, especially where the work of many agents must concur, so was it found in this. For some of those that should have gone in England fell off and would not go; other merchants and friends that had offered to adventure their moneys withdrew and pretended many excuses; some disliking they went not to Guiana; others again would adventure nothing except they went to Virginia. Some again (and those that were most relied on) fell in utter dislike with Virginia and would do nothing if they went thither. In the midst of these distractions, they of Leyden who had put off their estates and laid out their moneys were brought into a great strait, fearing what issue these things would come to.

The next quotation deals with the perennial problem of funding. Here Bradford is quoting a letter written by Robert Cushman, the man responsible for buying provisions for the Pilgrims' voyage. He writes from Dartmouth on 17 August 1620, desperately late in the year, months after the ships ought to have started.

And Mr. Martin, he said he never received no money on those conditions; he was not beholden to the merchants for a pin, they were bloodsuckers, and I know not what. Simple man, he indeed never made any conditions with the merchants, nor ever spake with them. But did all that money fly to Hampton, or was it his own? Who will go and lay out money so rashly and lavishly as he did, and never know how he comes by it or on what conditions? Secondly, I told him of the alteration long ago and he was content, but now he domineers and said I had betrayed them into the hands of slaves; he is not beholden to them, he can set out two ships himself to a voyage. When, good man? He hath but £ 50 in and if he should give up his accounts he would not have a penny left him, as I am persuaded. Friend, if ever we make a plantation, God works a miracle, especially considering how scant we shall be of victuals, and most of all ununited amongst ourselves and devoid of good tutors and regiment.

My last quotation describes the fear of disaster, as it appeared in the debate among the Pilgrims over their original decision to go to America.

Others again, out of their fears, objected against it and sought to divert from it; alleging many things, and those neither unreasonable nor improbable; as that it was a great design and subject to many inconceivable perils and dangers; as, besides the casualties of the sea (which none can be freed from), the length of the voyage was such as the weak bodies of women and other persons worn out with age and travail (as many of them were) could never be able to endure. And yet if they should, the miseries of the land which they should be exposed unto, would be too hard to be borne and likely, some or all of them together, to consume and utterly to ruinate them. For there they should be liable to famine and nakedness and the want, in a manner, of all things. The change of air, diet, and drinking of water would infect their bodies with sore sicknesses and grievous diseases. And also those which should escape or overcome these difficulties should yet be in continual danger of the savage people, who are cruel, barbarous and most treacherous, being most furious in their rage and merciless where they overcome; not being content only to kill and take away life, but delight to torment men in the most bloody manner that may be.

I could go on quoting Bradford for hours, but this is not the place to do so. What can we learn from him? We learn that the three devils of disunity, shortage of funds, and fear of the unknown are no strangers to humanity. They have always been with us and will always be with us whenever great adventures are contemplated. From Bradford we learn too how they are to be defeated. The Pilgrims used no technological magic to defeat them. The Pilgrims' victory demanded the full range of virtues of which human beings under stress are capable; toughness, courage, unselfishness, foresight, common sense, and good humor.

Bradford would have set at the head of this list the virtue he considered most important, a faith in Divine Providence.

I end this sermon on a note of disagreement with Bernal. Bernal believed that we shall defeat the Devil by means of a combination of socialist organization and applied psychology. I believe that our best defense will be to rely on the human qualities that have remained unchanged from Bradford's time to ours. If we are wise, we shall preserve intact these qualities of the human species through the centuries to come, and they will see us safely through the many crises of destiny that surely await us. But I will let Bernal have the last word. Bernal's last word is a question which William Bradford must often have pondered, but would not have known how to answer, as he watched the first generation of native-born New Englanders depart from the ways of their fathers.

We hold the future still timidly, but perceive it for the first time, as a function of our own action. Having seen it, are we to turn away from something that offends the very nature of our earliest desires, or is the recognition of our new powers sufficient to change those desires into the service of the future which they will have to bring about?

APPENDIX E

INFRARED OBSERVATIONS AND DYSON CIVILIZATIONS

MARTIN HARWIT
Center for Radiophysics and Space Research
Cornell University, Ithaca, New York

Some years ago, Sagan and Walker (1966) suggested searches for Dyson civilizations (1960) based on the infrared emission coming from the extended inhabited shell. For a shell maintained near 300°K, peak emission would be expected in the 10-micron range if the spectrum is approximately thermal.

Data from four types of observations are now available:

1. Humphreys, Strecker, and Ney (1971) have found that six luminous G supergiants have circumstellar emission shells whose spectra peak in the 10-micron region. The fractional emission in the long wavelength part of the spectrum, however, amounts to no more than a few percent. The authors believe that the infrared emission is produced by circumstellar silicate grains.

Other stellar objects, many of which were unearthed by the Neugebauer and Leighton (1969) "Two-Micron Sky Survey" are much fainter in the visual part of the spectrum but do emit strongly at infrared wavelengths. Many of these are currently believed to be late M supergiants, heavily obscured by interstellar dust.

2. A number of bright infrared objects are associated with HII regions. In many cases the HII region is very compact and shows heavy dust obscuration. No visual radiation is observed but strong thermal radio emission earmarks these objects. Apparently Ly-α radiation emitted by the ionizing central stars is absorbed by dust distributed in or around the HII region and the energy is reradiated in the infrared. This class of object is discussed by Ney and Allen (1969) and by Harper and Low (1971).

3. The total emission from galaxies might be considered to provide at least an upper limit on the number of Dyson civilizations. These upper limits, however, are very high for those galaxies bright enough to yield reliable data: M82, several Seyfert galaxies, etc. Such bright objects may emit more radiation in the infrared than in all other wavelength regions combined (Kleinmann and Low, 1970).

4. Background radiation measurements in our own galaxy may yield an upper limit to the total number of Dyson civilizations; but this limit is very high too. The diffuse 10-micron background radiation measured from rockets is probably dominated by thermal emission from interplanetary dust. This circumstance is likely to hold true even along directions far off the ecliptic plane. The minimum signals observed (Soifer, Houck, and Harwit, 1971) are comparable to and somewhat larger than the night sky brightness produced by integrated starlight, in the visual part of the spectrum.

In summary, infrared emission may be a recognition mark for Dyson civilizations, but other criteria will have to be formulated if a search is to distinguish these civilizations from naturally occuring infrared astronomical objects.

References

Dyson, F. J. (1960). Search for Artificial Stellar Sources of Infrared Radiation. *Science* 131:1667.

Harper, D. A., and F. J. Low (1971). Far Infrared Emission from HII Regions. *Astrophysical Journal* 165:L9–L14.

Humphreys, R. M., D. W. Strecker, and E. P. Ney (1971). High-Luminosity G Supergiants. *Astrophysical Journal* 167:L35–L40.

Kleinmann, D. E., and F. J. Low (1970). Observations of Infrared Galaxies. *Astrophysical Journal* 159:L165.

Neugebauer, G., and R. B. Leighton (1969). Two-Micron Sky Survey—A Preliminary Catalogue. NASA SP–3047. Washington, D.C.

Ney, E. P., and D. A. Allen (1960). The Infrared Sources in the Trapezium Region of M42. *Astrophysical Journal* 155:L193–L196.

Sagan, C., and R. G. Walker (1966). The Infrared Detectability of Dyson Civilizations. *Astrophysical Journal* 144:1216–1217.

Soifer, B. T., J. R. Houck, and M. Harwit (1971). Rocket Infrared Observations of the Interplanetary Medium. *Astrophysical Journal* 168:L73.

APPENDIX F

SEARCHING FOR GODot

JOSHUA LEDERBERG
Department of Genetics, School of Medicine
Stanford University, Stanford, California

I assume (1) the angels are still within the Galaxy, and (2) their average "progress" can best be estimated as equal to ours ± millennia.

Then a reasonable proportion will have reached the stage of interstellar (wisely "unmanned") travel—but not of being able to discriminate Earth from the multitude. (The UFO seers are incredibly geocentric in their conceits.)

Ergo: if there is a transmitter, it will be at the unique point in the Galaxy—as far as I can imagine, the *barycenter* is just that. This will be even more self-evident to that subset of angels who live a few hundred or thousand light years from the center. By the way, would they have more interstellar dust to use for rocket "fuel"?

Further I would favor looking for signals in the frequency domain, as compared with the advantages of time-averaging. The operational program is to look for unexpected line-structure, superimposed on the doppler-broadened lines of, say, the 21-cm

H microwave emission. Which we have good reason to do any-
how.

But if one takes such programs seriously, we must also con-
template the political imperatives of establishing terrestrial *radio
silence* vis à vis emissions that might be detected elsewhere.

APPENDIX G

TACHYON BIT RATES

MARTIN HARWIT
Center for Radiophysics and Space Research
Cornell University, Ithaca, New York

Whether or not tachyons could be used for communication between advanced civilizations remains a point of dispute. A number of causality arguments have suggested that tachyons may not even exist. However, until this issue is settled, it is worth asking what bit rate could be achieved with tachyon communication and how this compares with transmission making use of electromagnetic waves.

We assume that phase-space arguments determine the distinguishability of tachyons and that the number of distinguishable tachyons transmitted per unit time determines the bit rate. For a receiver with area A and solid receiving angle Ω, the volume from which tachyons are received per unit time is ANc, where N is the tachyon speed measured in units c, the speed of light. The momentum space volume occupied by these tachyons is $\Omega p^2 \, dp$, per mode of polarization. The number of distinguishable tachyons incident on the detector, in unit time (here referred to as bit rate) therefore would be

$$\left| \frac{ANc\Omega p^2 \, dp}{h^3} \right|.$$ (1)

We make use of the relativistic expression

$$\epsilon^2 = p^2 c^2 + m^2 c^4 = m^2 c^4 (1 - N^2)^{-1}$$ (2)

relating energy ϵ and rest mass m to momentum and velocity. This leads to the (imaginary) momentum value

$$p = \frac{N}{\sqrt{1 - N^2}} mc$$ (3)

and expression (1) obtained for a velocity range dNc reads

$$\left| \frac{A\Omega}{h^3} m^3 c^4 \frac{dN}{N^3} \right| \qquad \text{for } N \gg 1.$$ (4)

The corresponding expression for electromagnetic radiation is

$$\frac{A\Omega v^2 \, dv}{c^2},$$ (5)

where dv is the frequency of the radiation. If we take the frequency to be that of visible light and take m to be an electron mass, the tachyon bit rate is seen to be many magnitudes greater than the electromagnetic bit rate, as long as N remains less than about 10^7 and $dN/N \sim dv/v$. At that speed, the energy per tachyon would be about $10^{-7} mc^2$ corresponding to about 0.1 electron volt while the visual radiation would require a transmission energy about an order of magnitude higher.

If $N \sim 10^8$, the bit rate and energy expenditure per message is comparable to that for visible light, but communication across the universe can be achieved in times of the order of 100 years.

An increase in the tachyon velocity increases the transmission velocity, but decreases the bit rate. Messages with low information content can therefore be transmitted very rapidly.

Finally, one should note that there is a fundamental question involving the interrelation between tachyons and our definition of cosmic horizons. Clearly, current concepts such as particle, event, and absolute horizons have no fundamental meaning if tachyonic messages can be transmitted across horizons which exist only for photons and ordinary matter.

APPENDIX H

X-RAY PULSES FOR INTERSTELLAR COMMUNICATION

JAMES L. ELLIOT
Laboratory for Planetary Studies
Cornell University, Ithaca, New York

X rays are not appropriate as a means of transmitting a continuous stream of information because of their high quantum noise. But as a means of sending and receiving the all-important first bit, the announcement message, x rays have some strong advantages.

One advantage is that we have already used x rays to send out signals. These signals were sent in the course of several nuclear explosions. When a nuclear weapon explodes, about 70 percent of the energy released is in the form of kilovolt x rays, and this x-ray pulse is formed in less than a microsecond (Westervelt and Hoerlin, 1965). If the explosion occurs above 80 km, the x rays are not absorbed by the atmosphere and are free to propagate into space. The United States has performed at least five high-altitude tests of which two had yields in the megaton range (Westervelt and Hoerlin, 1965; Brown, Hess, and Van Allen, 1963; Van Allen, McIlwain, and Ludwig, 1959). Assuming that the Soviet Union has tested a similar number of high-altitude nuclear bombs, we can estimate that our planet has sent about ten powerful x-ray pulses into outer space.

Other advantages of using the x-ray pulse generated by a high altitude nuclear explosion to establish initial communications are (1) the pulse is short and will not be broadened or appreciably attenuated by propagation through the interstellar medium; (2) there are no stringent requirements on the receiver, since the pulse covers a broad x-ray spectrum; and (3) the x-ray flux involved is much larger than that of the sun, a local source of background x rays.

The quiet sun, in the microsecond time of a nuclear explosion, produces about 9×10^{17} ergs of x rays in the 1 to 10 kiloelectron volt range (Neupert, 1969). However the 1.4 megaton bomb exploded by the United States in the "Starfish" test yielded about 3×10^{22} ergs of 1 to 10 kiloelectron volt x rays. Even a large solar flare would emit, in a microsecond, less than one-hundredth of the x-ray energy emitted by the Starfish bomb. Also we must consider that the time variation of a solar flare is on a scale much larger than a microsecond.

There are two limiting factors in the detection of nuclear x-ray pulses at large distances. First we must receive enough photons in the detector to define a "pulse," and this number must be many times the statistical fluctuation in the background flux of cosmic x rays. The background flux is about 12 photons per square centimeter per second per steradian in the 1 to 10 kiloelectron volt range (Giacconi, Gursky, and Van Speybroeck, 1968).

Assuming that the fluctuations in the x-ray background follow Poisson statistics, we can derive a formula for the maximum distance d (in light years), at which an x-ray pulse from a nuclear explosion can be detected. We shall assume that the bomb has a yield of E megatons (1 megaton = 4×10^{22} ergs), and that 50 percent of this energy goes into 1 to 10 kiloelectron volt x rays in the form of a pulse lasting 1 microsecond. The mean energy of the x-ray photons produced is approximately equal to 2 kiloelectron volts. In addition we shall hypothesize a method of channeling a large fraction η of this x-ray pulse into a solid angle Ω_b. This would intensify the x-ray pulse emitted in certain directions at the expense of the ability to signal in all directions simultaneously.

To detect the x-ray pulse we can use a satellite-borne proportional counter that is 80 percent efficient for detecting x rays in the 1 to 10 kiloelectron volt region (Giacconi, Gursky, and Van Speybroeck, 1968). The solid angle viewed by the detector is Ω (steradians) and its area is A (square centimeters). To have less than 1 false alarm (cause by the x-ray background) per year we require the x-ray pulse to be at least 15 times the root mean square fluctuation in the number of background photons detected in one microsecond. These assumptions lead to the equation for d:

$$d = 2.9 \times 10^{-3} \left\{ EA \frac{4\pi\eta}{\Omega_b} \right\}^{1/2} (A\Omega)^{-1/4} \text{ light years.} \qquad (1)$$

In addition to the signal-to-noise criterion of equation (1) we shall also require that at least 15 x rays of the pulse be detected. This requirement leads to the inequality

$$d \leq 1.9 \times 10^{-4} \left\{ EA \frac{4\pi\eta}{\Omega_b} \right\}^{1/2} \text{ light years,} \qquad (2)$$

which can lower the value of d found by equation (1).

We can now calculate the distance at which the Starfish explosion could be observed by x-ray detectors now in use. A typical proportional counter, sensitive in the soft x-ray region, might have an area of 1000 square centimeters and a $5° \times 5°$ field of view. Using $E = 1.4$ megatons and assuming that the x-ray pulse is equally intense in all directions, we find that equation (2) gives $d \approx 400$ astronomical units—about ten times the radius of Pluto's orbit.

If the United States and the USSR pooled their nuclear stockpiles to produce a single large explosion (far from the earth!), the pulse could be detected at a considerable distance. We shall make a wild guess that each superpower has an arsenal of 10^4 megatons ($E = 2 \times 10^4$ megatons). Also we assume that the x-ray pulse can be concentrated into a conical beam of about 30° in angle with no loss of x rays. The detector will have $A = 10^4$ square centimeters and $\Omega = 0.08$ steradians. In this case equation (1) gives $d = 190$ light years—a respectable distance for signaling over a relatively large solid angle.

A search for interstellar signals, in the form of microsecond x-ray pulses, could be carried out in conjunction with x-ray astronomy observations. Since many galactic x-ray sources exhibit large flux variations on short time scales, millisecond time resolution is a common feature of most x-ray experiments. In principle it should not be difficult to extend the time resolution of these detection systems to times as short as a microsecond, and to monitor the time and arrival direction of all unusually large microsecond pulses. This microsecond pulse search could be in operation the same time as the detector was being used for other x-ray astronomy observations.

References

Brown, W. L., W. N. Hess, and J. A. Van Allen (1963). Collected Papers on the Artificial Radiation Belt from the July 9, 1962, Nuclear Detonation, Introduction. *Journal of Geophysical Research* 68.

Giacconi, R., H. Gursky, and L. P. Van Speybroeck (1968). Observational Techniques of X-Ray Astronomy, *Annual Review of Astronomy and Astrophysics* 6.

Neupert, W. M. (1969). X Rays from the Sun. *Annual Review of Astronomy and Astrophysics* 7.

Van Allen, J. A., C. E. McIlwain, and G. H. Ludwig (1959). Satellite Observations of Electrons Artificially Injected into the Geomagnetic Field. *Journal of Geophysical Research* 64.

Westervelt, D. R., and H. Hoerlin (1965). The Los Alamos Air Fluorescence Detection System. *Proceedings of the IEEE* 53.

AN ELEMENTARY GLOSSARY

Abiological
In the absence of life.

Angular momentum
A measure of rotational inertia; for example, the tendency of
a rotating object to continue rotating. For a point mass, the
angular momentum is its mass times its velocity times its
distance from the center of motion.

Anisotropic
Not isotropic; beamed in some particular direction or directions.

Astrometry
Precise measurement of the positions and motions of stars.

Astronomical unit (AU)
The distance of the earth from the sun: 93 million miles or 150
million kilometers.

Band pass
The range in frequencies accepted by a radio receiver or
transmitted by a radio transmitter. Band passes can be described
as wide band or narrow band.

Baryon charge conservation
An apparent law of nature that the net electrical charge on
heavy nuclear particles in any interaction is unchanged.

Bit
The minimum possible unit of information; essentially a yes or
no answer to a given question.

CETI
Acronym, invented for this meeting, for *C*ommunication with
*Extra*t*errestrial *I*ntelligence.

Civilization, Type I
A civilization able to use the equivalent of the present energy
output of terrestrial civilization for interstellar communication.

Civilization, Type II
A civilization able to use the equivalent of the energy output
of the sun for interstellar communication.

Civilization, Type III
A civilization able to use the equivalent of the energy output of
the Milky Way Galaxy for interstellar communication.

Cybernetics
The study of machine control processes; loosely, the design
of artificial intelligence.

Decimeter
One-tenth of a meter, a typical wavelength in radio astronomy.

Dispersion
In the propagation of electromagnetic waves in a medium,
the arrival of some frequencies of radiation before other
frequencies.

Doppler effect
The change of frequency of electromagnetic radiation due to
a relative motion along the line of sight between transmitter
and receiver.

Eccentricity
The departure of a planetary or other orbit from a circle.

Equation of state
The relationship connecting such properties as temperature, pressure, and density in a given material.

Erg
A small unit of energy, comparable to that of a 2 gram mass moving at a velocity of 1 centimeter per second.

ETI
Extraterrestrial Intelligence.

Exosphere
The outermost regions of a planetary atmosphere from which escape to space occurs.

Fourier analysis
The decomposition of a signal into a set of simple harmonic curves (sines and cosines) of appropriate amplitude and phase.

Flux
The amount of energy crossing some specific area in some specific time.

G-dwarfs
Stars roughly like the sun.

Gaussian distribution
A normal or bell-shaped distribution curve, common in statistics.

Gigahertz (GHz)
1 billion hertz.

Gravitational constant
A quantity expressing the strength of the gravitational attraction between two objects at a given distance and mass.

Hertz (Hz)
One cycle per second, a frequency unit in common use in
electrical engineering and radio astronomy.

Hydroxyl
A molecular fragment consisting of an oxygen and a hydrogen
atom.

Interferometer
A device that uses light interference to achieve high spatial
resolution.

Isotropic
The same in all directions.

°K
Temperatures on the absolute or Kelvin scale. Absolute 0 is
−273° Centigrade. The cosmic blackbody background is only a
few degrees above 0°K.

Kilohertz (kHz)
1000 hertz.

Kiloparsec
One thousand parsecs.

Kilowatt
One thousand watts.

Light year
The distance that light, traveling at 186 thousand miles per
second, travels in one year. It is almost 6 trillion miles or almost
10^{18} centimeters.

Lincos
A proposed language for interstellar discourse invented by the
Dutch mathematician Hans Freudenthal.

M-dwarfs
Most stars less massive and fainter than the sun.

Magnitude
A measure of the brightness of a star. Two stars five magnitudes different in brightness differ in their light output by a factor of one hundred.

Megahertz (MHz)
1 million hertz.

Megawatts
1 million watts.

Micron (μ or μm)
One-ten thousandth of a centimeter, a frequent unit in measuring the wavelength of infrared radiation.

Microsecond
One-millionth of a second.

Microwaves
Short radio waves.

Millisecond
One-thousandth of a second.

Mole
A mass in grams numerically equal to the molecular weight. One mole contains 6×10^{23} particles.

Nanosecond
One billionth of a second.

Nonzero sum games
Games where the net losses are not necessarily balanced by the net gains. For example, games where everybody wins or loses.

Omnidirectional
Isotropic.

Order of magnitude
A factor of 10. Thus three orders of magnitude is 1000 or 10^3.

Parsec

A distance of 3.26 light years.

Phased array

A collection of radio telescopes electronically connected so they function collectively.

Photons

The irreducible minimum components of electromagnetic radiation—light of any wavelength or frequency—ranging from gamma rays, X rays, ultraviolet light, and visible light, through infrared radiation and radiowaves.

Planck emission

The electromagnetic radiation emission spectrum of a perfect blackbody, where the emission is due to the temperature of the source. Real emission spectra of solid or gaseous objects have more complex spectra.

Pleistocene

A recent geological period, marked by glaciations, and beginning 2 to 3 million years ago.

Polymer

A long chain molecule, consisting of simpler building blocks called monomers. For example, proteins are polymers of amino acids and nucleic acids are polymers of nucleotides.

Polypeptides

Long chain polymers of amino acids, similar to proteins.

Pulsar

A rapidly rotating neutron star.

Quanta

See Photons.

Quasar

A "quasistellar radio source." In most, but not in all, theories a very energetic stage in the evolution of centers of galaxies.

Ramjet, interstellar

A proposed device which uses the matter of the interstellar medium as a working fuel and reaction mass for interstellar spaceflight.

Refractory

Involatile.

Signal-to-noise ratio

The amount of energy in the signal we are listening to divided by the amount of energy in sources of noise ("static") which are preventing us from hearing the signal. In most applications detection requires a signal-to-noise ratio much greater than 1.

Sinusoidal signal

A simple harmonic signal, a sine or cosine.

Spectrum

The arrangement of electromagnetic radiation according to its frequencies; an indication of the way material absorbs or emits electromagnetic radiation at different frequencies.

Superconductor

A material that permits electricity and heat to travel across it with almost perfect efficiency.

$3°K$ Cosmic blackbody radiation

The temperature of an object at equilibrium in space, first detected at radio frequencies. In most but not all theories, it is attributed to the remnant of the radiation from the primeval fireball out of which the universe arose.

Vapor pressure

A measure of the amount of a gas at a given temperature in equilibrium with its condensate; as, for example, the amount of water vapor over ice at a given temperature.

Watt

10 million ergs per second.

$>$

Greater than.

$<$

Less than.

$\sim$

Of the same order of magnitude as.

INDEX TO DISCUSSIONS BY PARTICIPANTS

SUBJECT INDEX